# THE TASTE OF WATER

**CRITICAL ENVIRONMENTS: NATURE, SCIENCE, AND POLITICS**

*The Critical Environments series publishes books that explore the political forms of life and the ecologies that emerge from histories of capitalism, militarism, racism, colonialism, and more.*

Edited by Julie Guthman and Rebecca Lave

1. *Flame and Fortune in the American West: Urban Development, Environmental Change, and the Great Oakland Hills Fire*, by Gregory L. Simon
2. *Germ Wars: The Politics of Microbes and America's Landscape of Fear*, by Melanie Armstrong
3. *Coral Whisperers: Scientists on the Brink*, by Irus Braverman
4. *Life without Lead: Contamination, Crisis, and Hope in Uruguay*, by Daniel Renfrew
5. *Unsettled Waters: Rights, Law, and Identity in the American West*, by Eric P. Perramond
6. *Wilted: Pathogens, Chemicals, and the Fragile Future of the Strawberry Industry*, by Julie Guthman
7. *Destination Anthropocene: Science and Tourism in The Bahamas*, by Amelia Moore
8. *Economic Poisoning: Industrial Waste and the Chemicalization of American Agriculture*, by Adam M. Romero
9. *Weighing the Future: Race, Science, and Pregnancy Trials in the Postgenomic Era*, by Natali Valdez
10. *Continent in Dust: Experiments in a Chinese Weather System*, by Jerry C. Zee
11. *Worlds of Green and Gray: Mineral Extraction as Ecological Practice*, by Sebastián Ureta and Patricio Flores
12. *The Fluvial Imagination: On Lesotho's Water-Export Economy*, by Colin Hoag
13. *The Low-Carbon Contradiction: Energy Transition, Geopolitics, and the Infrastructural State in Cuba*, by Gustav Cederlöf
14. *Unmaking the Bomb: Environmental Cleanup and the Politics of Impossibility*, by Shannon Cram
15. *The Taste of Water: Sensory Perception and the Making of an Industrialized Beverage*, by Christy Spackman

# THE TASTE OF WATER

## Sensory Perception and the Making of an Industrialized Beverage

CHRISTY SPACKMAN

UNIVERSITY OF CALIFORNIA PRESS

University of California Press
Oakland, California



Library of Congress Cataloging-in-Publication Data

Names: Spackman, Christy, author.
Title: The taste of water : sensory perception and the making of an industrialized beverage / Christy Spackman.
Description: Oakland, California : University of California Press, [2024] | Series: Critical environments: nature, science, and politics; 15 | Includes bibliographical references and index.
Identifiers: LCCN 2023023564 (print) | LCCN 2023023565 (ebook) | ISBN 9780520393547 (cloth) | ISBN 9780520393554 (paperback) | ISBN 9780520393561 (epub)
Subjects: LCSH: Water—Purification—Taste and odor control—United States—20th century. | Water—Purification—Taste and odor control—France—20th century. | Drinking water—Sensory evaluation—United States—20th century. | Drinking water—Sensory evaluation—France—20th century. | Municipal water supply—United States—20th century. | Municipal water supply—France—20th century.
Classification: LCC TD430 .S633 2024 (print) | LCC TD430 (ebook) | DDC 628.1/6—dc23/eng/20230608
LC record available at https://lccn.loc.gov/2023023564
LC ebook record available at https://lccn.loc.gov/2023023565

33 32 31 30 29 28 27 26 25 24
10 9 8 7 6 5 4 3 2 1

Chapter 1 is derived in part from Christy Spackman, "Just Noticeable: Erasing Place in Municipal Water Treatment in the US during the Interwar Period," *Journal of Historical Geography* 67 (January 2, 2020): 2–13. Chapter 2 is derived in part from Christy Spackman, "Perfumer, Chemist, Machine: Gas Chromatography and the Industrial Search to 'improve' Flavor," *Senses and Society* 13, no. 5 (2018): 41–59.

*For Bree,*
*Who nudged me at an inflection point*

# Contents

# Illustrations

# Acknowledgments

THIS BOOK came into being thanks to a constellation of mentors, funders, and collaborators. As others have aptly pointed out, sole authorship is in many ways a fiction.[1] I am especially indebted to Marisa Manheim for closely thinking, researching, and creating interventions with me around direct potable reuse over the past four years. While I have been interested in water reuse since 2015, Marisa is the one who suggested we look more closely at the move to turn water into beer as part of our work in the Sensory Labor(atory). It has become increasingly difficult to disentangle our thinking over the tenure of our collaboration; I have tried to mark throughout the footnotes and text of chapter 5 our explicit points of coproduction. However, ideas that once felt like mine increasingly seem to be the chimeric offspring of our collective work. Thank you, Marisa, for pushing me in new directions.

Four fellowships of varying sizes played an especially important role in bringing this book into being: an SSRC

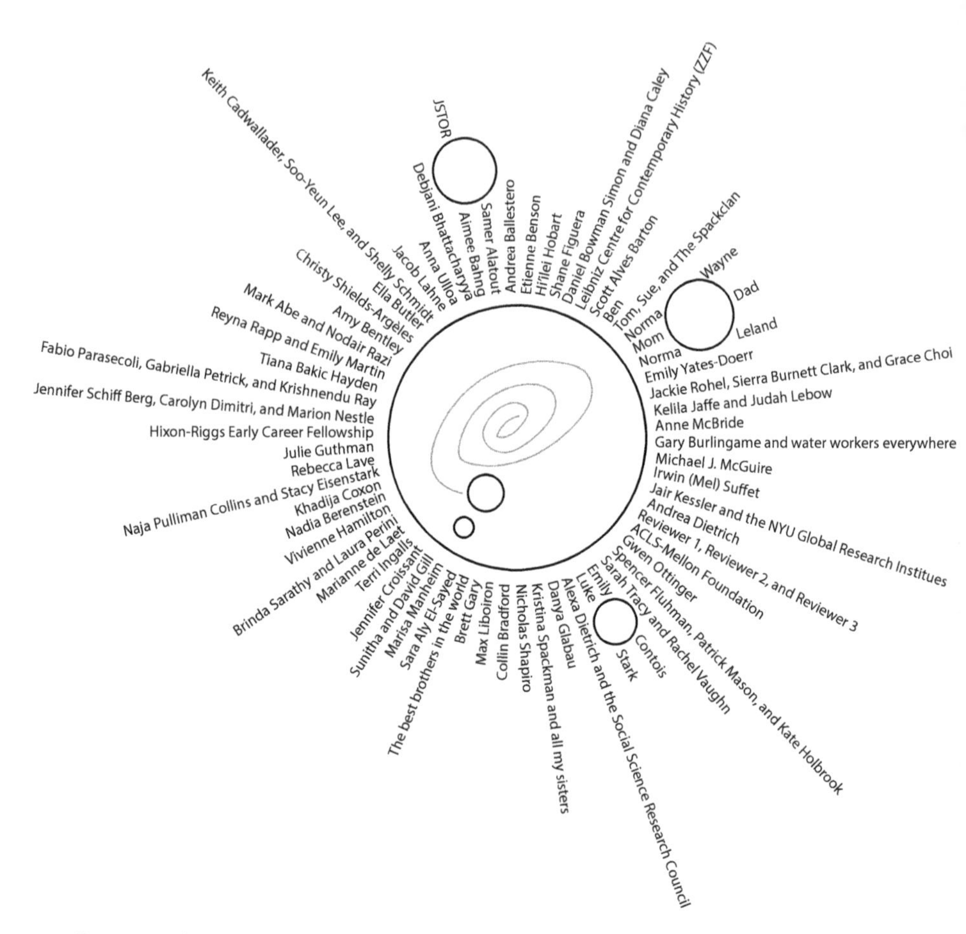


**Figure 1.** Author's patchwork constellation of mentors

dissertation proposal development fellowship put me on a new path under the mentorship of Caroline Jones and Peter Galison; a Mellon-ACLS dissertation completion fellowship let me take a break from teaching to survive so I could finish writing up my dissertation; the Hixon-Riggs Early Career Fellowship in Science and Technology Studies at Harvey Mudd College allowed me to strengthen my skills linking STS with food studies; and finally, a summer fellowship at the Leibniz Centre for Contemporary History facilitated book proposal development and completion of the initial manuscript.

I think it is easy to get swept up in the imagination of the perfect mentor, a single superstar who draws out road maps, gently guides you, offers critique, provides opportunities, and opens doors. I've instead found myself developing a theory of patchwork mentorship over my career. What is patchwork mentorship? It's the mentorship that one cobbles together, finding the necessary bits and pieces across a network of people at different career stages rather than a single individual. With that in mind, I've drawn on Melissa de Leon Mason's paper-pieced, machine- and hand-quilted work, *We Are All Made of Stars* (2021), to pattern my recognition of the different people and institutions who have mentored me through writing this book (see figure 1).[2] Thank you Melissa, for allowing me to transform your quilt into a patchwork of gratitude. An extra dose of appreciation to Andrea Ballestero, Etienne Benson, Julie Guthman, and my anonymous reviewers for taking on the entire manuscript and providing the sort of feedback everyone needs; to Gwen Ottinger and Ideas on Fire for developmental editing; and to Hi'ilei Hobart for talking me through an especially thorny rewrite challenge.

To my family, immediate and extended, words do not do my gratitude justice. Thank you for being along for my small part in "participating in the work of creation."[3]

# Abbreviations

| | |
|---|---|
| ADL | Arthur D. Little Company |
| AWWA | American Water Works Association |
| AZPWBC | Arizona Pure Water Brew Challenge |
| BHVP | Bibliothèque Historique de la Ville de Paris |
| CIRSEE | Centre International de Recherche sur L'Eau et l'Environnement |
| DPR | direct potable reuse |
| FPA | flavor profile analysis |
| GC | gas chromatography |
| GC-O | gas chromatography olfactometry |
| IAWPRC | International Association on Water Pollution Research and Control |
| MWD | Metropolitan Water District |
| OWBS | One Water Brewing Showcase |
| RO | reverse osmosis |
| TON | threshold odor number |

# INTRODUCTION

ON MAY 3, 2022, Chicago mayor Lori Lightfoot stood in the city's historic water tower and announced the launch of a new municipal water branding campaign: Chicagwa. "We are here today," Lightfoot told the gathered officials and members of the press, "to draw more attention to how we use our city's beloved crown jewel, Lake Michigan."[1] Run in association with National Drinking Water Week, the Chicagwa campaign used a limited-run set of canned water (with cans designed by local artists) and a cheeky ad campaign to promote the quality of Chicago's municipal drinking water. Bottled or straight from local taps, Chicagwa's "great drinking water" came from neither "an exotic island" nor a "fancy glacier." Instead, as the short film campaign narrated by urban historian Shermann "Dilla" Thomas pointed out, Chicagwa water came from right next door: the great lake "snuggled up" against the city's eastern border.[2] "Water is the reason Chicago even exists," Thomas noted as a mustachioed actor drank water, "it's pretty much Chicago's entire past and also its future. And we're sitting on a nearly endless supply of it, which the Department of Water Management will be turning into clean, refreshing drinking water long after our great-grandsons' great grandsons can grow their own thick mustaches."[3]

When I moved to Chicago's South Side neighborhood, Hyde Park, in 2001, I was surprised by how much I liked the drinking water. I liked it *despite* the fact that it came from the corner of Lake Michigan once infamous for its polluted waters teeming with wastes from the Chicago meatpacking and other industries. I grew up in a different Hyde Park—a small town of less than five thousand in northern Utah—and would happily describe the heavily mineralized, minimally treated, mountain spring water from my hometown as the ideal water to anyone who asked. Despite my taste for hard water, I really liked the water coming out of the tap in my little Chicago flat.

My family did not share my fondness for this new Hyde Park's water. Years later, my brother admitted that when he and his wife visited, they snuck bottled water into my apartment to drink on the sly. The water coming from my taps, he recalled, tasted a "little bit musty and dry, almost like it had de-oxygenated. . . . [I]t tasted like chlorinated lake water." Had we walked over to Lake Michigan and taken a sip of the raw lake water, we would have encountered a completely different beverage altogether.

The water coming out of taps in small towns like the one I grew up in, or in large metropolitan areas across the world like Chicago or Paris, tastes and smells fundamentally different from the raw water that enters municipal water systems. Someone—many someones—has done a lot of work over the past hundred or so years to manage the tastes and smells of water delivered throughout municipal water systems. Their work has "taught" tap water drinkers to expect water to taste a certain way: to expect, for example, that the water in the Hyde Park neighborhood in Chicago *could* and maybe even *should* taste like the water in Hyde Park, Utah. Indeed, many of the people I have talked with over the last decade describe the water they drink out of taps or bottles as "good" or "bad." Yet when I ask people how their water tastes, they often struggle to respond. Most say their preferred water tastes of nothing.

When I describe the water from my former Chicago neighborhood as good, and my brother responds that it is bad, we are both highlighting our personal tastes rather than some quantifiable quality of the water. The *personal* nature of such preferences makes them subjective. Policy makers and scientists in the twentieth century generally excluded matters of taste from regulatory systems due to the subjective nature of personal preferences.[4] Despite the subjectivity of sensory experiences, the people in charge of producing municipal water worked very hard over the twentieth century to figure out how to make water's tastes and smells fade into the background so that consumers could ignore or overlook its flavor. Making water taste like nothing is still one of their core goals. Their success has depended on the development of new forms of sensory and technical expertise. In fits and bursts, over the twentieth century waterworkers got better at communicating with each other about how to identify, treat, and manage unwanted tastes and smells in the water they produced. With each improvement of their skills, waterworkers made it increasingly easy for drinkers to not pay attention to the relationship between the water they drank and the natural and man-made environments it came from. As this book argues, that expertise put a wedge between how many individuals experience and understand the world surrounding them and how that environment actually is.[5]

This book is about the work that has gone into making drinking water taste relatively unremarkable in countries with well-developed municipal water infrastructures. It focuses on the development of new practices of sensory expertise over the twentieth century in the United States and France and investigates how that expertise has shaped the management of tastes and smells found in raw and treated drinking water. It asks what impact the changing types of sensory data available to everyday drinkers have had on how people with a range of different levels of expertise respond to the ingestible environment: the molecules, minerals, and materials that make up things we eat and drink. This book claims that the work of erasing tastes and smells in municipal water has altered awareness of the ways that the environment has been polluted, and in the process has come to shape the personal, political, and technological decisions shaping our environmental futures.

## PAYING ATTENTION TO SENSORY DATA

The types of sensory data we pay attention to shape what we sense. Similarly, what we sense shapes what we pay attention to. A waft of smoke or the rotten-egg smell from the sulfur-containing molecule mercaptan, which is added to natural gas, can catch and direct our attention toward the environment, but only as long as we are capable of perceiving these cues.[6] These little bits of perceptible data activate action. Smoke invites us to check the oven, look for a fire, or flee a building. In contrast, what we cannot taste or do not smell offers different lessons about the environment: erase the ability to smell, and the mercaptan causing that nasty rotten-egg odor will fail to signal that anything is wrong, sometimes with disastrous consequences.[7] Similarly, the perceptible data found in water, its tastes and smells, influences how individual drinkers and their neighbors, friends, and colleagues react to that water. For example, when the water coming out of taps remains unremarkable, day in and day out, it becomes easy to assume that everyone across the municipality, region, state, and beyond enjoys the same luxury. Such assumptions can get in the way of attending town council meetings. They can make it hard to support expensive new infrastructure projects. Sensory data, made imperceptible, paves a path toward inaction.

Nineteenth- and early twentieth-century urban industrialization resulted in an out-of-sight, out-of-mind approach to disposing of polluted waters, often to the dismay of downstream locales. As areas urbanized, physical environments were reconfigured in ways that prioritized urban dwellers' needs over rural water rights. Prioritization of wealthy urban inhabitants' desires for water over the needs of a city's poorer residents mirrored the geographical unevenness in access.[8] The World Health Organization (WHO)/United Nations International Children's Emergency Fund (UNICEF) Joint Monitoring Programme for Water Supply and Sanitation notes that approximately three-quarters of the world's population has access to a safely managed water source. While that portion may seem large, it also means that an estimated one out of every four people does not have access to such a source. Within the United States, 97 percent of the population in 2020 had access to safely managed water supplies, leaving approximately 9.89 million people who *should* have access to safe water without it.[9] In some cases, this lack of access is due to rural use of unregulated water sources such as homestead wells, whose safety depends on whether activities such as mining, fracking, or smelting have contaminated the water.[10] In other cases, lack of access is due to urban infrastructural decay.[11] Many still live in a world characterized by compromised water quality.[12]

At the same time, for people with access to what the US Environmental Protection Agency (EPA) defines as community water systems, the water coming out of their taps is often plentiful, relatively affordable, and generally of good—or good enough—quality.[13] This evaluation comes from the most readily available toolkit humans have for evaluating water quality: eyes, noses, and mouths—everyday sensors that indicate no need for worry. Water that lacks flavor, that provides refreshment, allows concerns about infrastructural failure or environmental degradation to fade away. This is water that invites drinkers to put their attention elsewhere.

Yet putting one's attention elsewhere carries risks to individuals, societies, and even the watery environments that sustain all life. A century of chemical innovation altered aquatic ecosystems: plastics are now found in rural freshwater streams, deep in remote ocean currents, and even in the Antarctic; pharmaceuticals and personal care products from sunscreen to shampoo appear in waterways; radionuclides from mining or nuclear weapons testing contaminate waters throughout the desert Southwest; and salts used to soften water, melt ice, or fertilize fields impair surface waters and wetlands.[14] At the same time, megadroughts and climate change threaten the viability of communities, be they along coasts or in arid regions.

With all of this in mind, you might ask why you should pay attention to water's aesthetic qualities—be it coming out of a tap or flowing into a municipal water treatment facility—when there are so many more pressing challenges around access and safety. This is a good question. It shares an assumption currently codified by regulatory structures in countries like the United States: that good tasting water is a luxury, while safe water is a right. It is a good question, also, because it highlights the divide defining whose expertise is allowed to matter in policy decisions and regulatory codes. Paying attention to the management of water's aesthetic qualities makes it possible to see that the lack of flavor in many drinking waters is not at all natural.[15] In writing about the taste of water, rather than just safety, I invite you to join me in taking seriously the role that sensory data can and has played in shaping how experts and everyday consumers govern environmental futures. In calling attention to the work of trying to erase smell and taste from water, I aim to stir the pot, to bring mouths and noses back into the work of thinking about our relationship with each other and our environment.

## INDUSTRIAL TERROIR

As soon as we start thinking of water as a food in addition to a substance necessary to life, a whole new world opens up. This is a world where water's tastes and smells *matter*. In prioritizing taste and smell, this book and its arguments walk a tricky line; as noted water scientists Irwin (Mel) Suffet and Joel Mallevialle point out, "Palatable waters aren't always potable."[16] For example, lead, with its ability to damage developing brains, is either undetectable or at especially high levels, tastes sweet.[17] Just because something tastes good does not mean it is safe.[18] By prioritizing water's perceptible qualities, I do not discount the significant public health gains made through twentieth- and twenty-first-century water treatment research. Rather, I aim to expand conversation in food studies, science and technology studies, and beyond to consider how technological innovations put in place to manage mundane moments of tasting and smelling link *and* unlink sensing, perceiving bodies and environments in ways that actively shape futures. Indeed, thinking of water as a foodstuff allows conversations about taste and smell to bubble up and sit alongside conversations about public health and safety—conversations that have dominated public-facing discussion about water production and circulation since governments realized that while stinky waters slowed economic growth, cholera-containing waters could entirely stop it.

Water rests uneasily in Western categorizations of food. It is an integral part of all foods. Like food, water is necessary for maintaining life. In calling water *food* I invoke all of food's other potential meanings beyond that of maintaining life. Food nourishes. Food is grown, harvested, prepared, husbanded, produced, slaughtered, cooked, eaten, wasted, and composted. In contrast, water is the substance whose presence *allows* food to exist. Water allows food to nourish bodies. By collapsing food and water, rather than continuing to hold them in separate but intertwined categories, I insist on prioritizing certain aspects of water over others: most obviously for this book, taste and smell, and to a lesser extent, texture. This insistence can, and perhaps ought to, be criticized for its very human-centeredness, given that water's presence and absence fundamentally shapes all life on earth. By insisting on water as food I center the fact that it will eventually interact with tasting, smelling, desiring bodies—many of which are human, and many of which are not.

The potable water flowing from city taps or sitting bottled on grocery store shelves is a highly industrialized product. When we start thinking about water as food, it becomes easier to question the assumption that the water coming out of taps, drinking fountains, and bottles naturally tastes and smells the way it does. It becomes easier to ask questions such as "Why doesn't my water taste more like the river I walk next to, or lake I fish in?" And perhaps more critically, "Why doesn't it bother me that they are different?" Once we start to pay attention to the tastes and smells found in water, it quickly becomes evident that there is a lot of work going on. This insight, though simple, is central to critically examining the work being done by engineers and innovators to address an anticipated future of water scarcity, and it provides a template for future work opening the boxes and bags circulating through the global food system.

Even as I insist in this book on collapsing water into food, I do not ignore that water is a substance constantly crossing uses. Only a small fraction—estimates generally land on about 3 percent—of the municipal water coming out of taps is used for eating and drinking. Water lubricates the workings of domestic and industrial sectors. If you recall high school chemistry, you may remember learning that water is capable of (eventually) dissolving or breaking down almost all of the things it comes into contact with. As such, the tastes and odors found in raw water mark water's travels: they reflect the microbes, minerals, soil, agricultural and industrial runoff, animals, and plants water encounters before it is treated and distributed. As explored throughout this book, especially in chapters 1 and 3, the molecules that perceptibly mark place have resisted and continue to resist technological taming. Instead, the tastes and smells found in the "cooked" water delivered through taps (and bottles) subtly signal where in nature a water came from as well as the human labor done to transform its taste.

The combination of what environmental historians refer to as a water's "biogeophysical genealogies" (the biological, geographical, and physical things and places a water has come into contact with) with active treatment processes aimed at managing perceptible markers of place results in a specific form of *terroir*: industrial terroir.[19] By using the term *industrial terroir,* I am riffing on an increasingly global mode of thinking about how food and place interact. Initially associated with French wine production, the term *terroir* emerged over the twentieth century as a way to frame how people relate to and think about the land foodstuffs come from and the work that goes into making those foods. In her examination of how the concept of terroir as taste of place expanded from France to the United States, Amy Trubek notes that nineteenth-century French speakers used terroir as an agricultural term that referred to the earth from which food came.[20] Nineteenth-century French folks primarily understood *goût de terroir* as foods that "tasted of the earth" they grew in, and secondarily understood *goût de terroir* as reflecting the labor practices, values, and production approaches used in producing foods.[21] For example, champagne became champagne in large part due to producers' efforts in the Champagne region of France to define their production methods and legally link those practices to the specific region and its soils. Producers drew on this self-imposed constraint in aggressive, external-facing advertising campaigns that promoted champagne not just as a beverage, but as a beverage with terroir.[22]

When advertisers and promoters of local foods use the term terroir, they are closely welding together ideas about the locality of how and where a food is produced and the larger cultural and economic values underlying its production.[23] Similarly, when producers use the idea of terroir to promote the tasty qualities of their wine, pork, cheese, maple syrup, tequila, or tea, they are using a form of terroir characterized by the taste of place *as made available* through labor practices designed to maximize the connection between taste and the place where a food is produced. Terroir closely links taste, place, and production practices throughout the food chain.

In using the term industrial terroir, I highlight how industrial food production seeks to divorce taste from place through technological, regulatory, and expert practices of *making unavailable* the sensory qualities that mark place-based uniqueness. Water producers working within the ideals of industrial terroir (even if they do not refer to it as such) aim to minimize and mask place-based uniqueness; in the case of water, the tastes minimized are not only the local tastes of earth, rocks, plants, and animals, but also the unique tastes of place caused by industrial pollution.

The creation of industrial terroir depends on the development of expert practices of working with the senses, something Jacob Lahne and I suggest can be categorized as sensory labor.[24] Sensory labor happens when observations about perceptible molecules found in the ingestible environment are transformed into data that can be used to shape individual or institutional decisions. Experts and everyday folks practice sensory labor. For experts, that work often occurs in field sites or laboratories with the objective of turning sensing into data that can circulate away from the bodies that did the sensing. Yet sensory labor is going on all the time. For example, each time you or I decide to purchase a "new and improved!" version of a familiar product, we compare the new experience with our memories of the previous product. If it aligns, or is better, we will purchase again. If not, we may decide to put our money elsewhere. As people perform sensory labor, they *produce value* for themselves and for others.[25]

While it may seem obvious that people have drawn on their sensing capabilities to navigate and measure the environment for all of human existence, the practice of turning human sensing into "data" is recent. Indeed, the idea of data itself is relatively new: what can become and count as data keeps shifting, even if the rhetorical framing of data as a "given" thing out there in the world remains.[26] The twentieth century witnessed a radical reconceptualization of sensory information. Chapter 2 shows that as researchers interested in sensing took advantage of new instruments like gas chromatographs, the way they understood tastes and smells changed. No longer were tastes and smells just experiences; rather they were caused by specific, identifiable molecules: tastants and odorants. With appropriate instruments and skills, scientists and technicians could measure and manipulate those molecules. They could even mobilize tastants and odorants as specific forms of data that people could sense (become aware of), perceive (become aware of and consciously recognize or categorize), and pay attention to.[27] Scientific and technological innovations allowed experts to use the knowledge gained through instruments to produce standardized taste profiles that everyday eaters perceived as the same, or at least within a range of sameness, be the item purchased in Chicago, Illinois, Phoenix, Arizona, or even in a town in another country.[28] Making sensory experiences into data facilitated the emergence of the increasingly uniform, industrialized terroir of municipal water. In the process, it set in place an unofficially held, but nonetheless powerful, set of understandings around the value of tastes and smells found in drinking water. To track who has access to specific sensory cues about the environment requires following the development of techniques for identifying and responding to tastes and smells.

## NOT-KNOWING WATER

One summer while conducting research in France, I came across a booth hosted by the City of Paris's municipal water provider, Eau de Paris, at the farmer's market in my neighborhood. One of the volunteers handed me a colorful map of the city split into different sections (see figure 2).

Each section, the volunteer told me, drew its water from a different source. One area south of the Seine got its water from an underground

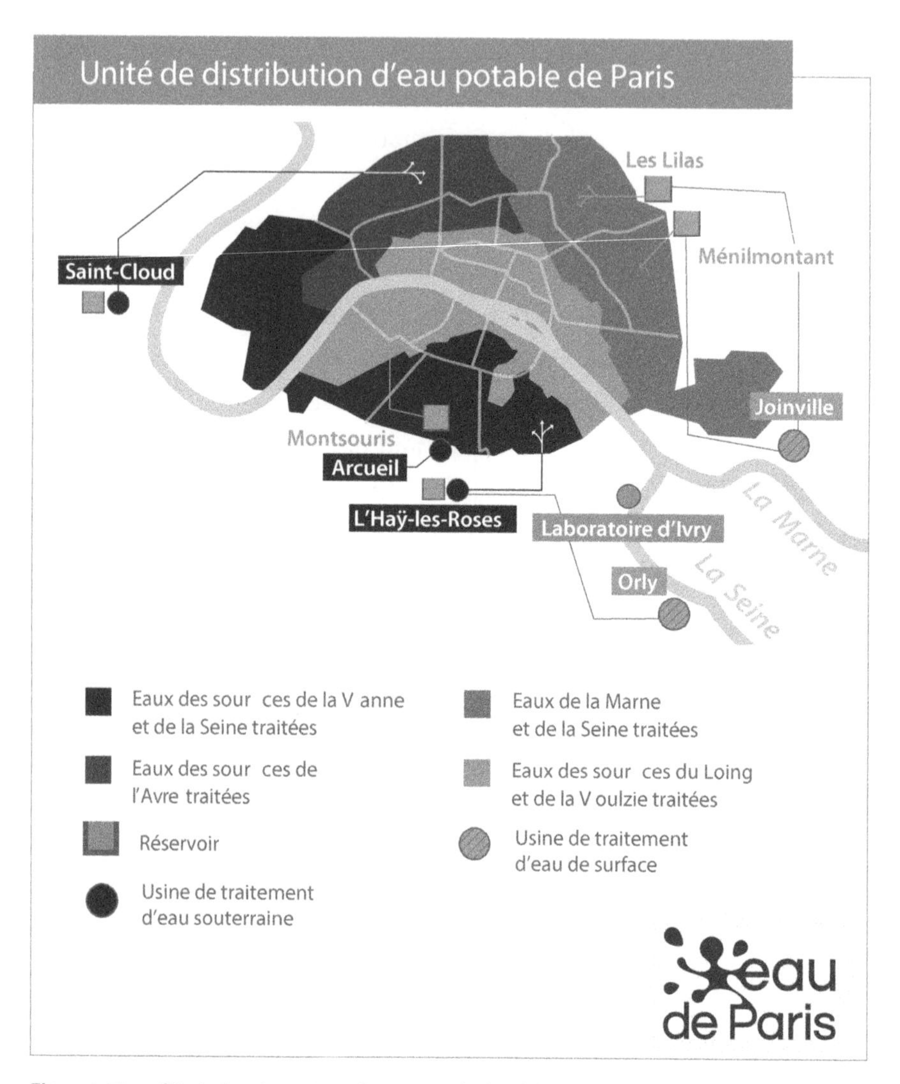


**Figure 2.** Map of Paris showing origin of water supplied in different neighborhoods. Courtesy of Eau de Paris.

aquifer. The region I lived in received water from a nearby river. All, the volunteer pointed out, had been treated for safety; all had essentially the same mineral content and flavor. Eau de Paris even offered beautiful glass water carafes for sale reiterating this message—each carafe had a unique design reflecting the arrondissement on one side and an identical "nutrition label" on the other outlining the mineral content of the water.

Without that colorful map and helpful volunteer, Parisian water would have remained a uniform product for me. I had drunk tap water at different places all around the city. Yet my mouth, nose, and eyes had found nothing remarkable to pay attention to. My combined senses did not notice that the water I drank in the library came from a different source than the water in my small apartment—indeed, from sources so different that had someone handed me a glass of the raw water from each I would have easily distinguished between the two. My senses *could not* alert me to those differences, because the differences that did exist were so small that they fell below my threshold of perception. I did not know something, because I could not sense it. In not being able to sense the differences, I could not perceive anything to pay attention to.

The study of how people come to not know things is called *agnotology*.[29] For example, governments and companies use security clearance and trade secrets to limit who can access information; Peter Galison estimates that "the classified universe is five to ten times larger" than what is found in our libraries.[30] Social customs can similarly limit what scientists pay attention to. Londa Schiebinger shows that colonial scientists collecting medicinal plants in the Caribbean chose to not bring back knowledge to Europe about the abortion-inducing powers of the peacock flower, despite being aware that pregnant slaves used it to avoid bringing children into slavery.[31] Dominant cultural values also influence what histories are recorded. Kevin Dawson points out that many Western histories are "physically and mentally landlocked" and as a result have contributed to overlooking the rich underwater cultural lives of enslaved Africans in the new world.[32] Built infrastructure also plays a role in shaping what can be perceived. As Emily Thompson demonstrates, changing architectural and building practices over the twentieth century made some sounds more perceptible, while other sounds were dampened and destroyed through the use of building materials like acoustic tiles and insulation.[33] Social, cultural, and physical infrastructures shape what things are unavailable to the senses and thus are moved outside of everyday realms of perception and attention.

As the preceding cases highlight, not knowing is unevenly distributed among people. Those with security clearances can access information that everyday people cannot, but that access may well change as people switch jobs or security priorities shift. British colonial scientists—and the people they interacted with—in the Caribbean were aware of the peacock flower's medicinal potential. Yet British scientists chose to not circulate that knowledge in a new context because it did not align with how they understood the world. Enslaved Africans challenged racial hierarchies through their aquatic prowess, a fact that more recent social discourses entirely overlook. We see in these cases different aspects of agnotology: how one comes to not know something is shaped by accepted processes of learning to understand the world (epistemology). It is also influenced by how groups actually understand the world itself to exist and function (ontology). What you do not know depends on your situation. It depends on historical processes. Not knowing comes about in different ways.

Processes of creating ignorance are political. That means they depend on a specific politics around how knowledge is produced, organized, and circulated. For example, Scott Frickel and M. Bess Vincent point out that after Hurricane Katrina, testing protocols used by the EPA and Louisiana Department of Environmental Quality only measured certain chemicals and certain sites, a process that overlooked other chemicals and other sites. "Tests will do only what they are designed to do, and nothing more," Frickel and Vincent note.[34] As such, the scientifically "appropriate" approach did not sample soil from sites that were historically home to industrial production, nor did they—due to legal restrictions—test private property.[35] By limiting what chemicals scientists tested for, and where those tests were done, testing allowed politicians and experts to declare the whole city safe, even if portions were not. At the same time, the tests themselves limit what can be known, a process that Michelle Murphy labels as "regimes of perception"—the governance processes that shape what can even be known.[36]

Regimes of perception not only shape what can be known, they also shape values. Max Liboiron points out that the approaches to regulating pollution developed over the twentieth century—which permit pollutants to be dispersed into the environment as long as they fall under certain historically defined thresholds—codify an assumption that *it is okay* for bodies to be burdened with the need to assimilate some level of pollution.[37] That codification, Liboiron argues, makes it easy to assume "that's just how it is" rather than take action to undo the "bad relations of a scientific theory that allow some amount of pollution to occur."[38] Once systems of perception become naturalized, it can be hard to recognize that the practices we use to make knowledge could be otherwise.

Not knowing includes practices of conceptualizing relationships between place, inhabitants' bodies, and pollution. When used by a dominant culture such practices can, as Hi'ilei Hobart argues, erase important historical or religious aspects of a place. Hobart points out that "discourses of absence" portraying Mauna Kea mountaintop in Hawai'i as "a place without humans, spirituality, nation, or even atmosphere" allowed settlers to develop the mountaintop despite native Hawaiian opposition.[39] This erasure, like the physiological burden on bodies who ingest pollutants that fall below legally or instrumentally recognizable limits, enacts its own form of violence.[40] Taken together, Frickel's, Vincent's, Liboiron's, and Hobart's work demonstrates that not knowing can inculcate and perpetuate harmful practices across generations.[41]

When it comes to foods, not knowing is also unevenly distributed. The notorious complexity of food systems, which draw together local, national, and transnational legal structures, values, weather patterns, monetary systems, and tastes (and that is just the beginning!), means that different actors in the food system have different levels of knowledge about the same thing. For example, the sensory information gleaned when one bites into a slice of mass-produced pie depends entirely on where one is situated in the food system, as well as one's specific relationship to the product. Everyday eaters rarely reflect on such complexity when tasting a food.[42] Yet for the food and sensory scientists charged with making certain the pie's flavor survives from the factory to someone's plate, a single bite of that same pie tells stories about what did or did not happen during transportation. As discussed in the following section, honing in on the not knowing that comes about through management of smells and tastes requires an analytical approach that goes beyond epistemology and ontology. It calls for an approach that can account for bodies with different levels of expertise engaging in similar actions with similar, sensed data points.

## INFLECTION POINTS

Smells and tastes disappear quickly. They resist historical analysis.[43] This makes tracking their absence doubly challenging: most studies of absence focus on objects of concern that are removed from sight or hearing, a focus that itself reflects a Western hierarchy prioritizing sight and hearing over other forms of sensorial experience.[44] To date, the technologies for inscribing visual and (to a lesser extent) audio information into mediums that can endure is much more advanced than those that seek to inscribe taste or smell—as made evident by the relatively short shelf life of even the most processed foods or finest of perfumes when read against the long arc of history. Given the ephemerality of tastes and smells, how might one get a handle on the work over the twentieth century of removing troubling forms of sensory data, and its environmental implications?

To investigate the work of removing troubling forms of sensory data, in this book I identify and analyze different "inflection points" in innovations related to water management. I draw on a methodological approach from the fields of sociology of knowledge and science and technology studies that is used to evaluate how ignorance comes into being and circulates. I apply insights from that methodology that suggest paying attention to time, the size of information, the scale at which information circulates, and intentionality can help identify when moments of ignorance are being created. I further add place as a category. Each of these categories is not only good for identifying moments of ignorance; it can also identify important moments when an innovation in water management has had an impact on what sorts of sensory knowledge about the environment are available to the general public.

Sociologist Jennifer Croissant, who is particularly interested in the types of not knowing that develop around athletic injury, suggests a possible pathway forward. Croissant notes that in addition to looking for the structures that shape what we know, it can also be useful to pay attention to time (chronicity), size of information (granularity), scale, and intentionality in evaluating how ignorance comes into being, is used, and circulates.[45] I add another dimension, place, to the categories to pay attention to: water picks up tastes and smells as it travels, and much of the work of managing industrial terroir focuses on removing the markers of *where* and *what* a water has been in contact with.

Unlike major infrastructural decisions such as whether to build a dam or find a new water source, the changing modes of dealing with taste and smell over the twentieth century happened through small, experimental adjustments in caring for water. These adjustments, made in different places at different times, rarely came about in response to one-time disasters or extreme swings. Instead, the bulk of these adjustments came about because of nuances—inflections—in the taste or odor of raw or treated water that resulted in consumer complaints. Frequently, the most frustrating of these taste and odor inflections showed up at the threshold where imperception transitioned to perception: some people, but not everyone, noticed undesirable smells or tastes. For the engineers and others tasked with making municipal water, figuring out how to identify, communicate about, and act on the thresholds where perception happened played a key role in defining whose sensory experiences were allowed legitimacy, and under which circumstances.

Like the subtle inflection of a voice indicating surprise or disgust in a conversation, inflections in taste or smell contribute to, derail, or shift the possible decision pathways that individuals and institutions choose. For substances that humans can perceive, sensory inflection points—or to use the technical term, just noticeable differences (JNDs)—are activated when molecules interact with the receptors in smelling, tasting bodies. The presence of sensory inflection points can trigger feedback loops embedded in individuals and institutions. Yet it is easy to overlook that something like water, which can appear entirely natural, is part of much larger political, environmental, and technological contexts. It is even easier as a scholar, or activist, or drinker to overlook what things are going on at the point where just noticeability bleeds into imperception, especially since those boundaries change from person to person at different times. The following chapters show that, as engineers and chemists worked to develop tools for identifying and naming the molecules responsible for triggering feedback loops, they slowly improved their ability to transform raw water into water that the majority of consumers perceive as being like the water found in nature, just safer.

In this book I follow the creation and diffusion of new tools, knowledge structures, and networks of expertise that came to define late twentieth- and early twenty-first-century approaches to managing tastes and odors. The efforts to manage tastes and odors happened in fits and spurts, rather than continuously. Such efforts often occurred in response to specific environmental challenges. Technological, scientific, fiscal, and political constraints shaped how individual utilities and networks of experts responded to smelly or tasty defects. As a result, the unrolling story of the work done to address unpalatable water shared in the chapters that follow may initially appear disconnected in both time and place. The techniques developed by engineers, chemists, and marketing/communications teams in Chicago, Illinois; Philadelphia, Pennsylvania; Los Angeles, California; Paris, France; and Scottsdale, Arizona, over the twentieth and into the twenty-first centuries are linked by a shared desire to change how water consumers perceive the quality of their water. These cities have different water sources. They face different environmental challenges. Chicago draws from Lake Michigan's vast body of fresh water; Philadelphia from the Schuylkill River; Los Angeles is fed by a mix of local groundwater and imported surface water from other regions in California as well as the Colorado River, as well as recycled water; Paris uses a mix of local groundwater and surface waters; while Scottsdale's water comes from the Colorado River, the Salt and Verde Rivers, and groundwater that is recharged with recycled water. Despite their differences, all of these cities have struggled with varying water quality linked to changes in the built and natural environments. Yet the general population remains largely unaware of the struggles required to maintain a relatively uniform drinking water. That imperception is what the innovation in measuring and understanding taste and smell aimed to achieve.

Traditional studies of how innovations diffuse into society—from adoption of new pesticides or medicines to refusals to use improved cookstoves by home cooks in India or automatic washing machines in Canada—aim to evaluate the effectiveness of communication about a new idea.[46] Diffusion, innovation researcher Everett Rogers argues, is "a kind of social change" that depends on its ability to circulate among networks of decision makers.[47] Scholars interested in how innovations diffuse into society point out that if one graphs adoption of an innovation over time, the "path usually follows an S-curve."[48] Mathematicians call the point at which an S-curve changes directions an inflection

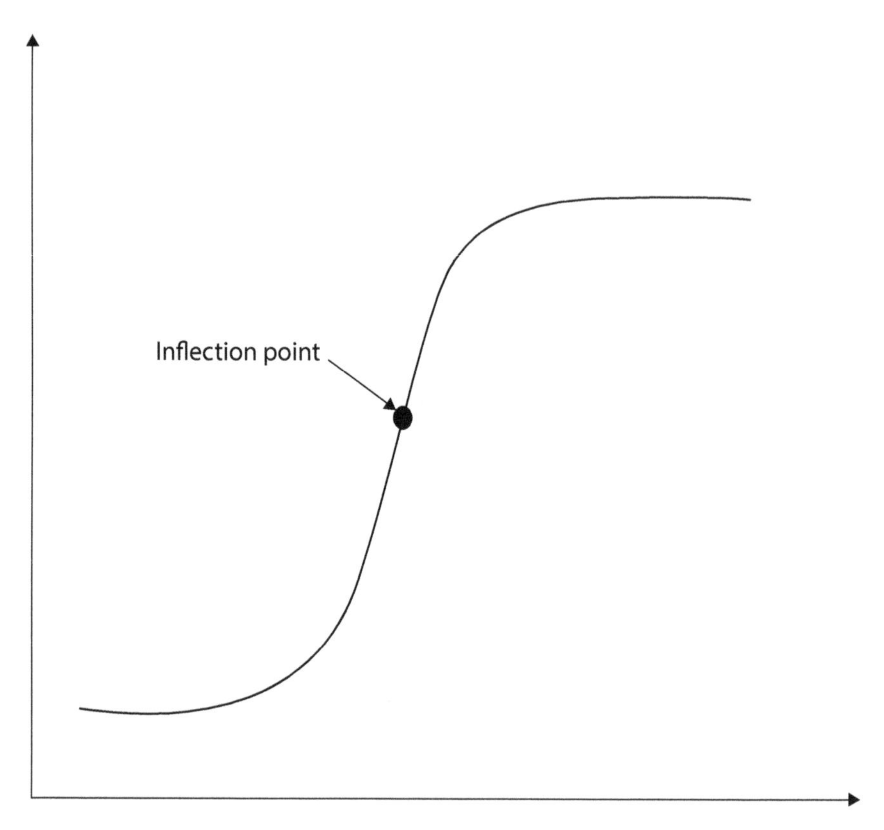


**Figure 3.** Inflection point on S-shaped curve. Drawing by author.

point; specifically, an inflection point happens when the rate of change of some variable shifts from positive to negative, or vice versa (see figure 3). In short, inflection points indicate a critical moment of change from the (mathematical) world as it has been to a new (mathematical) world in the making. The colloquial understanding of a change in perception that happens when someone just notices a difference is related to the mathematical way of describing inflection points. These curves help innovation researchers model how an innovation is taken up and used.[49] If you zoom in on an inflection point in a graph, that moment of change practically disappears. Zoom back out, and it becomes apparent that over time those minute changes happening at inflection points have significant impact. For scholars interested in why a technological innovation succeeded, the inflection point marks a key set of decisions that significantly shaped future success. Inflection points thus mark barely perceptible moments when new worlds come into being.

The perception of minute differences, that is, physiological inflection points, depends on a body's sensory capacities. For researchers, the variability between people's sensory capacities makes turning insights produced by different bodies into data tricky. In fact, Western philosophers in the eighteenth century such as Kant, Hume, Shaftesbury, and others argued that sensations of smell and taste were irretrievably subjective and fell outside of the measurement practices coming to define scientific practice.[50] By the late nineteenth century a small group of researchers in Europe and the United States began actively suggesting otherwise. These researchers, who called themselves psychophysicists, built on the work of German physicist-philosopher Gustav Fechner to develop ways of understanding and measuring embodied experiences. Like Fechner, psychophysicists were interested in the moment when a change in stimulus would be "just noticeable."[51] How much additional weight had to be added to a load before someone would notice? How much must a sound change before it woke up a sleeping person? The answers to these questions proved surprising: it takes a lot of weight to make a heavy load feel heavier, whereas an ant walking across someone's hand can entirely derail a conversation.[52] No matter when or how perception happens, once an inflection is perceived, it invites action.

For business strategists, inflection points mark the moment when a range of future possibilities begin to coalesce toward a future inevitability. This makes the ability to understand and identify inflection points especially useful for business strategists as they work to forecast a technology's success or failure, an approach popularized by Intel's Andy Grove. Grove situated what he termed strategic inflection points as the transition period during which a business significantly shifts how it functions. "The irony is that, even though these are the most cataclysmic changes that a business can undertake, more often than not these changes are missed," Grove argues. "You think nothing is going on because the curve is very smooth, the centre [*sic*] of the hurricane has no wind and you miss the whole thing until you are descending the curve."[53] Strategic inflection points often only appear in hindsight. Those who study innovations and their impact covet the ability to identify strategic inflection points: if one could recognize a strategic inflection point when it is happening, one could harness the change as it occurs, modify it, perhaps even change the innovation's directory into something that moves a business in desired, rather than unintended, directions. Inflection points retrospectively mark important strategic moments.

I use the ideas embodied by the inflection point—of physiological experiences that invite action and set in place the possibility of new futures—as a methodological entrance for examining how innovations inspired by an idea about what water should taste like have circulated through expert networks. I am interested in how inflection points have and might shape water governance; as such I organize each chapter around an inflection point in the governance of the taste of water. My approach differs from that of traditional studies of how innovations diffuse: looking for inflection points in the work of managing water production, rather than simply tracking how knowledge about managing tastes and smells in water diffused, allows me to attend to whose ways of knowing are allowed legitimacy, and when. In fact, conversations about *how* we don't know what we don't know remain incomplete as long as they overlook the ways that information normally transmitted by taste, touch/texture, and smell is made unavailable. By examining inflection points in action in consuming bodies, scientific research, and policy making, I open avenues for tracking not-yet-discovered vulnerabilities waiting in the moment when the trajectory of a system changes direction between past choices and future events. Inflection points facilitate examining how innovations move across fields.

Inflection points highlight how innovations can get enrolled in current efforts to shape how bodies in the future will sense.[54] Inflection points thus track the changing impact of expert sensory practices. In focusing on the inflection point, rather than the diffusion of innovations, I prioritize linking the processes producing environmental conditions to sensing bodies. In the next section I discuss how paying attention to time, scale, intentionality, granularity, and place enable identifying sensory inflection points.

## Identifying Inflection Points through Time, Scale, Intentionality, Granularity, and Place

Water used to be a very different beverage. Sanitarian George Whipple, writing about the value of pure water in 1907, characterized drinking waters found in New England as having a moderate amount of color and significant cloudiness. In contrast, people from the Midwest, "where all the streams are muddy," most often objected to unknown colors rather than color in general. Overall, he pointed out, most people could accept a small amount of cloudiness produced by small particles of clay. But the majority, he noted, rejected water with coarse sediment.[55]

Whipple's observations came at the cusp of a major shift in water provisioning: within a year of the publication of his book *The Value of Pure Water*, John Leal and William Corbin, who managed the water supply for Jersey City, New Jersey, would conduct the first large-scale chlorination of public water in the United States.[56] Chlorination produced new environmental conditions: up to that point, waterworkers primarily relied on sedimentation (letting sediment settle out of water) and various types of filtration to treat water before sending it out through the (often) newly built mains and pipes of municipalities.

To understand how innovations like chlorination have limited—and will continue to limit—the sensory capacities of bodies to perceive the environment, I pay attention to time, scale, intentionality, granularity, and place. For example, over time, adoption of chlorination not only reduced the bacterial loads in water; it fundamentally altered the "aesthetic" characteristics of water coming out of taps in ways that sedimentation and filtration had not. It removed color and turbidity.[57] Chlorination introduced new tastes and odors while simultaneously destroying others. As different municipalities adopted the practice of adding chlorine, the regional characteristics of water faded. Water increasingly became, as Ivan Illich suggests, an industrial beverage.[58] For example, childhood experiences with public swimming pools make the scent of chlorination easily recognizable for many. Yet most people living in places with trusted water treatment facilities have forgotten how different raw waters coming into treatment plants, as well as treated water coming out of taps or bottles, can be.

Chlorination did more than reduce the regional characteristics of water. In the early to mid-twentieth century, many producers found that despite the public health benefits of chlorination, its lingering odor left them open to criticism. Chlorine interacted with naturally occurring compounds in surprising ways: it killed living organisms in raw water, which sometimes caused an uptick in musty tastes and odors. When chlorine came in contact with phenol-containing substances found in tar-coated pipes or introduced by industrial dumping in waterways, an intensely medicinal taste ensued.[59] Tastes and odors found in drinking water acted as data points, marking relationships between new, public-health based ideals, economic growth, and scientific authority.

Have you ever tried to describe a new flavor to a friend? If so, you know that communicating about tastes and smells is really difficult. I think the cactus flowers blooming on a March night outside my back door smell a bit like rain. But not quite. Unless you've stuck your nose inside one, my writing that it smells like rain is not particularly helpful—a scratch and sniff sticker would be a much better communication tool! Researchers and regulators interested in taste and smell have developed different approaches to talking about smell over the twentieth century.[60] As demonstrated in the next chapter, early to mid-twentieth-century researchers in Chicago drew insights from their mid-western colleagues and from the emerging science of psychophysics to create a numerical system for talking about smells. The numerical systems did not prove as helpful as hoped; chapters 2 and 3 explore how it took the development of new tools for linking specific molecules with specific sensations *and* the rise of new types of environmental challenges to generate enough of a push/pull for the creation of a new system for talking about tastes and smells. By building systems for talking about sensory experiences, researchers helped knowledge about expert sensation circulate between different locations. The changing methods of producing knowledge about tastes and smells introduced multiple *scales*, from micro to macro, of not knowing.

Differences between how bodies and instruments perceive, and how knowledge about sensation travel, are at the heart of what Gary Burlingame, the former director of Philadelphia Water's Bureau of Laboratory Services, and I refer to as sensory politics. We define sensory politics as institutional or regulatory practices that allow or exclude sensory knowledge from participating in systems of action.[61] Most everyday folks rely on individual experience, published reports, and word of mouth to evaluate water quality. In contrast, experts primarily evaluate water quality by using instruments to measure the presence or absence of chemicals or microbes; after all, instrumental measurements are what allow water utilities to demonstrate that they are in compliance with contemporary regulatory systems. These two different modes of knowing can result in a disjuncture between an expert's evaluation of a water (it's safe!) and that of a consumer (ewwwwww, this smells gross, can it possibly be safe?).[62]

Some of the dangers found in water are perceptible: leaking of uncharacterized chemicals used for making clean coal in Charleston, West Virginia, in January 2014 caused water throughout the distribution system to smell like black licorice.[63] Other dangers are harder to perceive, only directly detected through testing: these include lead leaching into the water in Flint, Michigan (2014–), Washington, DC (2001–), and Newark, New Jersey (2016–); cyanotoxins caused by harmful algae blooms in Lake Erie; and radionuclides seeping into groundwater from uranium mining tailings on Diné (Navajo) land.[64] Further complicating detection, current regulatory structures do not require testing for all of those substances. In some cases, perceptible proxies like smelly or discolored water, possible indicators of pipeline decay, can allow people to identify dangers in the absence of testing. Yet as the recent cases of lead leaching into pipes in Flint and Newark indicate, while perceptible proxies raised residents' concerns, officials overlooked or discounted residents' suspicions of those proxies as irrelevant aesthetic issues. When someone says they smell or taste something in the water, they are, at least according to how we currently understand these sensations, indicating that specific molecules are perceptibly present in the environment. Sometimes those molecules are not a problem all. But in some cases *a lot* is at stake when someone says, "I smelled something. I tasted something," and experts respond by categorizing people's aesthetic concerns as irrelevant or irrational.

As demonstrated across the book, controlling the granularity of the molecular information available offered waterworkers in the twentieth century the ability to make a nonverbal argument about their capability and authority to drinkers. The presence or absence of sensory cues can signal to drinkers—who include politicians, voters, and bill payers—differing modes of intentionality on the part of water providers; when water smells, tastes, or looks "wrong," consumers often interpret those sensory cues as signs of neglect or disinvestment. As municipal water providers addressed treatment challenges, and as regulators and public health officers responded to changing understandings of safety and danger, the importance of drinking water's tastiness faded into a secondary position. Increasingly, access to good tasting water became a privilege, unevenly bestowed by geography, historical legacies of protected watersheds, or regional rates of economic growth.

Everyone's ability to smell or taste varies. Training, health (as the rise in anosmia associated with COVID-19 demonstrates!), age, hormone levels, and other things shape perceptual ability.[65] Sometimes people's mouths and noses are much more sensitive than the most sophisticated analytical instruments; while someone might be able to perceive minute amounts of something, it can be *really* hard to prove that what they sensed actually was present.[66] To further complicate things, different tastants and odorants have different thresholds at which they can be perceived. In other words, the information packets of knowledge available to sensing bodies or measuring instruments can be really, really small—they have a granularity both amenable to and resistant to instrumentation, especially because our understanding of how taste and smell work is still in its early stages.

As municipal water providers worked to manage off-tastes and smells, and as regulators and public health officers responded to changing understandings of safety and danger, they shifted how sensory data became known and circulated. The different, place-based experiences with tastes and smells found in water in the past have shaped our present experiences. Paying attention to how the intersection of bodies, technologies, businesses, and geographic locations create inflection points facilitates tracing current efforts by scientists, technologists, politicians, and businesses to mobilize new approaches to managing water. New water management approaches, such as water recycling, are often framed around anticipated sensory futures linked to specific places, their environments, and future sensing bodies who have not yet come into being. Time, scale, intentionality, granularity, and place help identify sensory inflection points. Focusing on sensory inflection points highlights the extent to which our relationship with the environment, shaped through our senses, has been shifted by the work of national and international networks of experts dedicated to the work of producing safe and good-enough water.

## BOOK MAP

This book follows how innovations in managing tastes and smells found in water were developed and circulated across the twentieth and into the twenty-first centuries, first through expert networks and then to the larger public. Each chapter moves to a different location to track how time, scale, intentionality, granularity, and place have shaped the work of removing perceptible cues from water. To explore how experts have worked together to transform the taste of water over the twentieth century, I here start in the urbanized present, where municipal water is widely available but often replaced by bottled water. Rather than the well water my father grew up with in rural Idaho, most people in the United States where I live have access to municipally treated water.[67] The drive to remake the taste of water happened in tandem with the increasing urbanization that my parents and grandparents lived through.[68] As a result, the stories told here are primarily stories of people working in large municipalities. This is an important point: as Martin Melosi notes, the water needs of large municipalities differ from those of small towns and have shaped how public health officials, engineers, and city leaders worked together to build "sanitary" cities.[69] Urbanization influences water provisioning by further separating drinkers from the source of their water.

The increasing distance from the places and modes of production came about in part because municipal waterworkers in the early twentieth century prioritized making water *safe*. Engineers and sanitarians in US cities like Chicago understood safe water to be water free from dangerous bacteria or toxicants, a definition informed by emerging regulatory structures that sought to limit the transfer of waterborne diseases along railways.[70] For engineers and sanitarians, unwanted flavors did not necessarily make water unsafe, unlike the views held by experts a century earlier. For water producers in Chicago before World War II, the work of defining safety with relationship to unwanted flavors proved especially challenging. Unlike their counterparts in New York City who had access to a watershed fed by mountain snowmelt, Chicago's water supply came from surface waters that also served as a dumping ground for the region's industries. Chicken feathers visibly contaminated portions of the lake. Sometimes those feathers even came out of people's taps. Similarly, when the wind blew from the south where steel plants used coke and charcoal to heat metal, phenol's sweet, tarry odor not only wafted across the beaches of Chicago's South Side, it also appeared in homes miles from the lakeshore. Chlorine, the tool most readily available to make water safe from pathogens, unfortunately interacted with chemicals like phenol to produce disgusting, medicinal flavors. Making water safe butted up against waterworkers' desires to transform water into something that resembled a deeply held, centuries-old aesthetic ideal for drinking water: clear, cool, abundant, potable, *and* palatable. Their desire seems entirely reasonable on any hot day.

What palatable water is varies from person to person. It varies from location to location. Despite regional and individual differences, those tasked with producing Chicago's municipal water in the early twentieth century understood that while some waters tasted or smelled good, others did not. Producers wanted to remove the perceptible markers of place—the chicken feathers from slaughterhouses or phenol from oil refineries—that kept undermining their efforts to produce water that aesthetically aligned with their modernist ideals. Unfortunately, the nascent nature of water treatment complicated waterworkers' efforts to draw on internal and external expertise. How could workers in Chicago evaluate whether a treatment introduced in Philadelphia would be appropriate in Chicago, or whether a treatment developed in Chicago would work elsewhere? Chapter 1 examines how workers began experimenting with new approaches that would allow them to standardize their individual embodied experiences. A group working on Chicago's South Side was especially active in this process, developing, testing, and refining a new sensory measurement tool, the threshold odor number (TON). The TON helped improve waterworkers' ability to evaluate the effectiveness of their own treatment processes by allowing them to transform smells into numbers. These new formulations of sensory labor situated waterworkers like cooks: people who take a raw ingredient and then process it into something new. In the case of water, that something new was a product that would ideally be unremarkable: a product with industrial terroir.

The aquatic chefs in the treatment kitchens of municipal water plants did achieve some level of success in decreasing off-tastes and smells in water before the 1950s. But their efforts remained incomplete. Workers in Chicago and elsewhere reported that although they could tell something was present, they lacked the tools for identifying what, exactly, *was* present. Their struggles reflect a larger concern in the food industry: How can one identify and get rid of unwanted sensory cues without first understanding what those sensory cues are? Chapter 2 investigates a series of developments in analytical chemistry, food science, and sensory science from the 1950s to the early 1980s. Findings from new analytical technologies began reshaping how researchers understood taste and odor. With the help of instruments, researchers began breaking down complex mixtures (be they crude oil, foods, or even toothpaste) into their individual parts and isolating the parts of the mixture that contributed to its overall flavor. As researchers worked to associate particular molecules with particular sensations, they also sought to identify the thresholds at which something became perceptible and to develop a vocabulary for communicating about sensation. In the process, a new way of understanding taste and smell emerged: flavor became a mixture of tastants and odorants, understood as little molecular blocks of data amenable to manipulation. Identifying flavor's molecular nature opened doors for the creation of industrial terroir, be it the work of product developers in standardizing flavors or of waterworkers seeking to isolate and treat unwanted tastes or smells from the edible environment.

Like the oil chemists, quality control specialists, and food scientists who set the stage for quantitatively understanding flavor, waterworkers sought to better understand the smelly or tasty molecules found in their products. They too drew on this new understanding of flavor as molecular data to help them better control raw water's aesthetic excesses. Many of those excesses came from more than a century of weakly regulated industrial innovation. Environmental shocks including chemical dumping, phosphate-rich agricultural runoff, and pipeline leaks contributed to an increased public awareness of the environment's finite ability to absorb and process the by-products of human ingenuity and human greed during the second half of the twentieth century. Highly visible manifestations of pollution, such as when the Cuyahoga River caught fire in 1969, helped mobilize awareness and action.[71] To address concerns about the quality of raw water in the 1970s and 1980s, engineers and other water professionals reached across national lines to form new, transnational research networks to improve their capacity to address decreasing water quality. Chapter 3 examines how experts in the United States and France, dissatisfied with the effectiveness of the TON in helping them address late twentieth-century aesthetic challenges, collaborated to develop and diffuse technologies for doing sensory labor across geographic and linguistic lines. Unwanted flavors that leaked through treatment and out to the population perceptibly marked the decreasing quality of the raw water supply. In response, members of the French-US research network drew on, and then remixed, practices and tools from the wine world. Like wine producers during this era, waterworkers understood their product as carrying *terroir*, the taste of the place where it was produced. Unlike wine researchers and producers—whose tools were increasingly being mobilized to democratize wine tasting for everyone—waterworkers developed tools aimed at experts tasked with removing terroir. The centerpiece of their effort was the creation of a water flavor wheel linking the tastes or odors that could be present in raw water with their potential molecular causes. The water flavor wheel created a shortcut. It allowed waterworkers to use their noses or mouths to quickly make treatment decisions around a range of known causes of undesirable tastes or smells. It also signaled the desire of engineers and chemists to make the many markers of environmental pollution fade away into a non-noticeable, unremarkable nothing; in short, the water flavor wheel marked the absences central to successfully making industrial terroir.

These past inflection points bring us to the near present. As bottled water increased in popularity across the United States and elsewhere in the late 1990s and early 2000s, it shouldered aside its municipal competitors, coming to stand in for a glass of water as the sign of hospitality.[72] In venues such as sports arenas and even at some public schools, vending machines replaced drinking fountains.[73] For producers of the bottles of water sold in vending machines, grocery stores, and restaurants, reeducation of consumer taste became a central part of the competition to replace the drinking fountain. As environmental concerns made bottled water more suspect, municipal water producers also adopted taste education. Chapter 4 traces how bottled and municipal water producers in France in the early twenty-first century worked to train consumers to drink and evaluate water like wine through performative tastings in the corporate boardroom and the streets of Paris. By training people to performatively use their senses to pay attention to water's tastes and smells, bottled and municipal water producers invited the senses to become the core agents defining which aspects of water deserve attention: they centered taste and smell as water's most important characteristics. Processes of creating TONs, molecular descriptions of flavor, and water flavor wheels contributed to erasing flavor from municipal water. But teaching people to taste water like wine acts in a similar way to all of the other processes discussed here: it directs attention toward certain things, often at the expense of others.

Industrial terroir aims to remove perceptible traces of environmental variability, to direct attention toward standardized quality rather than a food's specific place-based origins. As population growth and climate change combine to shift the qualities and quantities of water available in regions around the world, consumers in countries with well-developed water infrastructure find themselves at yet another inflection point. Chapter 5 investigates how technological innovations combined with limited water supply have made the unpalatable idea of using wastewater as drinking water popular in policy discussions and with municipal utilities facing water shortages. In water scares regions many utility directors, policy makers, and regional planners see wastewater as a potential new source of water capable of addressing anticipated water shortages. Despite the established nature of technologies allowing water reuse, many policy makers and consumers continue to express hesitancy about using wastewater as a source for drinking water. To address these concerns, proponents of direct potable reuse (DPR) have created new spaces for sensory labor to occur, spaces that seek to act in the present on future sensing bodies. By partnering with craft beer brewers as water ambassadors, proponents of DPR water actively demonstrate the power of industrial production techniques to transform yucky presents into yummy futures. Unlike the industrial terroir of the twentieth century, DPR presents a possible future in which the molecular markers of bodies, time, and place disappear.[74] This creates a blank slate—a gustatory *terra nullius*—ready to be made into whatever locally or nationally identified flavor profile the utility wishes (or can afford) to recreate, and as such, to be made to represent any place or time.[75]

# 1

# INDUSTRIAL TERROIR

ON DECEMBER 28, 1927, demand for bottled water outstripped supplies on Chicago's South Side. A front-page article in the *Chicago Daily Tribune* the following day reported the water flowing from taps was so bitter that residents refused to drink it. The city's chief engineer, H. H. Gerstein, disagreed, insisting that the water offered "no danger whatsoever." After all, he had ordered a special chlorination step to ensure the water's safety. A few days later, waterworkers sailed out into Lake Michigan to sample water at different locations. Their goal? To pinpoint the origin of the bitter taste plaguing South Side consumers. The "obnoxious tastes and gross sewage pollution" responsible for the complaints, the team concluded, were due to a southerly wind moving phenol-contaminated waters and sewage northward from the Calumet River and Indiana Harbor into Lake Michigan. The contaminants undermined the safety and palatability of water flowing into homes across Chicago's south side.[1]

Unfortunately for residents, this experience was repeated a few years later. Over a two-week period from late December 1932 to early January 1933, phenolic wastes from newly built oil refineries in the Calumet area resulted in "innumerable complaints of chloro-phenol tastes in public water supplies."[2] Complaints came from as far away as Highland Park, Illinois—almost forty miles north of the initial discharge!

The phenol-contaminated waters were caused by wastes from the very industries that helped transform Chicago into the nation's crossroads.[3] Gerstein's report notes that "wastes from coke and charcoal byproducts plants, from creosoting works and gas plants, and oily wastes from oil fields and oil refineries" contributed to the degradation in the city's water quality.[4] Coke and charcoal by-products point to the steel industry—itself made possible due to easily accessed shipping routes linking the iron deposits of the northern shores of the Great Lakes to the coal mines of the Midwest—while creosoting works highlight the production of railroad ties linking East and West.[5] Many of these industries were located at Chicago's southeastern edge and continued into Indiana, a spatial distribution understood to limit impact on the city proper, as long as winds, storms, or ice flows did not interfere with typical circulation patterns.[6] But winds, storms, and ice flows on Lake Michigan did interfere with typical circulation patterns, an environmental reality that resulted in a constantly changing raw water supply for the city. In the early decades of the 1900s, this meant the quality of water coming out of Chicago taps, its turbidity, odor, and taste, occasionally brought evidence of the burgeoning twentieth-century industrial ecosystem into the intimate spaces of home, eliciting complaints in the process.

Chicago's waterworkers were neither alone nor the first in encountering phenol-contaminated waters. A switch in the methods of coke production after World War I from air-polluting beehive ovens to water-polluting by-product coke ovens meant that areas with high steel production—primarily in the Ohio River basin, Detroit, Cleveland,

and the Chicago region—saw a substantial increase in phenol-polluted waters.[7] Phenol can come from both natural causes and human activity. Sources of phenol include oils produced as organic matter decays, waste from motorized vehicles, creosote linings in pipes, and of course, compounds contained in sewage.[8] This "new trade waste" resulted in a "disagreeable medicinal taste," especially when combined with chlorine.[9] Even very small amounts of phenol can be tasted. The chemical proved especially troublesome in cold weather, a marked contrast to many other taste and odor challenges.

When the medicinal-smelling and tasting water reappeared in late 1932 and again in early 1933, consumers complained loudly. Its smelly impact on the more than four million people living in Illinois and Indiana, as revealed in newspapers and professional journals, offers a pungent glimpse of the role that water's shifting perceptible qualities played in everyday life. Chicago's inhabitants and the city employees tasked with managing water safety were not the only ones who noticed when water quality differed from what was considered normal. For example, a Minneapolis city chemist reported in fall of 1928 that Minneapolis found itself facing a significant "menace": the "overnight" transformation of the city's water filters to "green as a lawn." A strong moldy odor accompanied the alarming visual transformation of the filters. Unfortunately, the normal treatment approach—large doses of alum—failed to do anything. Soon, "the very disagreeable moldy odor spread throughout the plant," and from there out into the entire city. To the mortification of city officials and waterworkers alike, the distasteful episode coincided with a meeting of the American Public Health Association. "The remarks that were passed about our water were no credit to any water works official" grumbled the city chemist to his fellow professionals.[10] Water's perceptible qualities shaped how inhabitants and visitors saw a city.

As the twentieth century unfolded, managing the taste and smell of water became increasingly important to city officials and employees working with municipal water supplies. This proved challenging, as local raw water sources often also served as sites for waste disposal. "Nearly every city is now under the need of using second hand water," opined noted public health expert Dr. William A. Evans in 1933 in his syndicated column, "How to Keep Well." His opinion came in response to a (hypothetical) reader's question: "Can water be made safe?" Known both for his work as health editor of the *Chicago Daily Tribune* and for his role as Chicago's commissioner of health, Evans's opinions reached a wide audience across the United States.[11] Evans lauded the success of twentieth-century water treatment in drastically reducing the threat of typhoid. He pointed out that chemists, engineers, and bacteriologists had "got down on their marrowbones" and innovated: "The waterworks plant of every city at present is an old clothes shop where a worn-out article is taken in hand, brushed, de-greased, de-nearly everything else, taken to pieces, and made into a bright, brand new suit of clothes." Evans's awareness of the challenge posed by trade wastes tempered his optimism about the previous decade's technological innovations: "Few communities have access to first class sources of water. There are too many factories and too many people for that. Cities are forced to take muddy water, sewage-polluted water, or water into which factories have poured all kinds of waste, and purify it as well as that can be done."[12] Purifying water, as well as it could be done, included not just preventing typhoid. It also meant navigating between different modes of making and evaluating quality water.

This chapter examines how city employees in Chicago and their professional counterparts in the municipal water industry transitioned from a focus on changing the physical landscape to trying to alter the chemical makeup of the raw water entering treatment facilities. As city engineers researched different strategies for managing tastes and smells in drinking water during the period between World War I and World War II, they worked to create the possibility for a future in which unwanted, unfamiliar, or disagreeable tastes and odors no longer threatened a city's reputation or a waterworker's job. By intervening in the perceptible sensory qualities of Chicago's water, city engineers set in place an inflection point at which waterworkers became shepherds of the industrial influences on the tastes and smells of Lake Michigan's raw water. Even if considered irrelevant to public health experts, at least as codified by regulatory structure in the United States, the ability to measure and alter the smells, tastes, and turbidity of water being delivered set in place an inflection point in a larger sensory rhetoric about what exactly water was, as well as who was most aware of its qualities.

## QUALITY WATER

Water producers in the late 1920s and early 1930s had a tricky job: the definition of what exactly constituted quality water had changed since the turn of the century. Up until discovery of the link between microbes and disease in the late nineteenth century, drinking water was considered "of good quality and potable when it is fresh, clear, without odor, when its taste is minimal, that it is neither disagreeable, nor insipid, nor salty, nor sweet; when it contains little foreign matter; when it contains a sufficient amount of dissolved air; when it dissolves soap without forming a scum and it perfectly cooks vegetables."[13] Most folks could perceive quality by looking at, smelling, or tasting their water.[14] As public health officials, physicians, politicians, and eventually everyday folks became aware of links between invisible microbes and disease, the work of evaluating quality transformed into a skill set no longer readily available to everyone. Instead, fields like biology and chemistry started to officially displace older definitions of quality and potability.

Making "quality" foodstuffs—be they the water discussed here or teas, wines, coffee, cheeses, breads, and so on—is a process that brings together a range of people, institutions, regulatory structures, and cultural practices. As Sarah Besky argues in her examination of what makes quality tea, quality is not simply a state of something. Rather, it emerges from relationships of power that shape how a product comes to be what it is, from the materiality of products themselves, and from the markets that shape and define sensorial expectations.[15] For waterworkers, changes in what made water "quality" went beyond definitions. Water has always reflected the places it came from. However, the "chlorine revolution," characterized by the new practice of dosing municipal water with chlorine-producing substances like hypochlorites and ferrochlore (a mix of iron salts and chloride of lime), meant that many urban drinkers soon began receiving water with tastes, smells, and even colors that widely varied from the water on well-fed farms and homesteads.[16] Massive infrastructure projects in metropolitan areas during the late nineteenth and early twentieth centuries further widened the divide between the perceptible qualities of water encountered by urban and rural drinkers. As early twentieth-century efforts to make water "pure"—itself an index for quality—were codified into standardized practices in municipal water plants, professionals found themselves occasionally at odds with consumer opinion; for example, operators in the early 1920s largely agreed that the relatively new approach of chlorination was "here to stay" regardless of consumer complaints.[17]

Over the late nineteenth and early twentieth centuries, the definition of what *good water qualities* were (e.g., palatability) and what constituted *quality water* (e.g., potability) diverged. The emerging regulatory structures defining quality, specifically the 1914 Bacteriological Standards for Drinking Water and its 1925 revision, set aside qualities of clarity, minimal odor, and minimal taste in favor of reducing bacteriological, and eventually chemical, loads to levels understood as safe—that is, to levels low enough that the public was protected from bacteriological diseases. That was good. Significantly fewer people got sick. Despite these changes, what exactly constituted quality water depended on one's viewpoint. For consumers, quality water was not only safe, it had good sensory qualities. In contrast, for water producers, quality water was first and foremost safe water.

## THE WATER MAKERS

Making quality water in the late nineteenth and early twentieth centuries was an act of emerging standardization not unlike writing a recipe. Standardizing water proved challenging: Chicago waterworkers knew Lake Michigan's surface water differed from groundwater found in Florida, for example.[18] To address concerns like differences in raw water sources, those working with water professionalized into the American Water Works Association (AWWA) in 1881. Its founding members, twenty-four men hailing from Kansas, Kentucky, Illinois, Indiana, Iowa, Missouri, and Tennessee, formed the association to connect the engineers, chemists, sanitarians, and technicians tasked with managing water.[19] They envisioned that regional meetings and a trade publication, the *Journal*, would facilitate communication about their techniques and approaches to water treatment.

Initially, the association grew slowly. In 1900 it reported only 350 members. By 1910 membership had almost tripled, and by 1914 the association had set up regional sections and established the *Journal*.[20] At first only published quarterly, by 1925 the *Journal* appeared monthly, publishing conference papers and responses, as well as technical and section news. Formation of committees to investigate pressing problems and to push for standardization closely followed the implementation of regional sections.[21] In short, the association sought to mitigate the problem, voiced in the 1915 presidential address to the association, that "when we examine in detail the annual reports of water works plants, all over the country, we find ourselves absolutely unable by comparative study of accounts to formulate any satisfactory comparison in almost any matter pertaining to them."[22] Facing qualitatively and quantitatively different natural and fiscal resources, by 1924 approximately 2,000 waterworkers across the United States had joined the AWWA to, as a colleague would put it twenty-two years later, "share experiences and work out common problems for the benefit of all."[23]

While all agreed that taste and odor problems were an ongoing challenge in the world of water treatment, these accounts—and the many others that peppered the pages of the *Journal*—also agreed that the public they served expected to receive water relatively free of unfamiliar tastes, odors, or colors. One engineer who regularly consulted with water providers wrote in 1934 that whereas consumers once expected an occasional unpleasant taste or odor from their municipal supplier, they now saw "no reason why any water supply, except under the most unusual circumstances, should have a disagreeable taste."[24] This expectation was a radical change from previous decades: St. Louis residents were once accustomed to drinking brown water, and residents in New England to drinking cloudy water; others stored their water before use so particles could settle out.[25] Professionalization went hand in hand with efforts to standardize water quality by providing the vehicle for waterworkers to communicate the results of experiments to minimize the variation in raw water's different qualities.

## INDUSTRIAL TERROIR

For the professionals tasked with providing Chicago with quality water, industrial activity in the region proved a notable threat. Located at the foot of Lake Michigan, Chicago had access to the largest body of fresh water in the United States. Lake Michigan connected Chicago to eastern ports, a critical part of a growing land- and water-based trade route that moved goods produced in the western and midwestern United States to hungry East Coast markets.[26] The lake provided plentiful fresh water for fighting fires and hydrating bodies. Its waters also served as a readily available industrial lubricant.

Water enables a range of industrial activities, such as corn refining, paper making, textile production, and steel manufacturing. Industrial wastes added chemical and biological burdens to US waterways.[27] For example, the Dow Chemical Company's main plant in Midland, Michigan, used almost all of the available water from the Tittabawassee River during low flow summer seasons before discharging waste-containing water back into the river.[28] This meant that in the summer the entire river essentially ran *through* the plant rather than by it. Dow began research into how to "recover or destroy" phenolic compounds in dilute solutions (a model system for the dilution of phenolic compounds in the river water) in 1927, but it took another decade before a treatment plant was placed in operation. In the interim, Dow resorted to holding "strong phenolic waste" (low pH) in storage ponds during winter months before diluting those wastes with river water and, after 1937, subjecting the wastes to settling and filtration.[29] Waterways were a critical piece in the nineteenth- and twentieth- century "industrial ecosystems" connecting natural resources with human industrial efforts.[30]

Although boosters saw Lake Michigan as a nearly infinite reservoir of clean water, waterworkers in Chicago and elsewhere along the lake in the nineteenth and early twentieth centuries found its waters of uneven quality. Anyone standing on the lake's shore near the mouth of the Chicago River could see that sewage and industrial waste contaminated the water.[31] To address this, in the mid-nineteenth century Chicago engineers began to build intake pipes farther from the shoreline. A two-mile-long tunnel out into the lake to bring waters in to shore, completed in 1867, was initially considered far enough away from the city that the water coming in would remain clean.[32] Population growth and industrial expansion undid that hope in the ensuing decades as even more waste drained into Lake Michigan's waters. Indeed, the wastes spilling from sewers, slaughter and packing houses, and ships into rivers, canals, and streams feeding into Lake Michigan were readily visible.[33] The lake's waters linked together the individual and industrial metabolisms of the city.

In a move documented by Chicago's boosters and detractors as a turning point in the history of US water pollution management, at the end of the nineteenth century city engineers reversed the flow of the Chicago River and opened the Chicago Sanitary and Ship Canal (1900), sending the "cesspool of filth" away from, rather than toward, the city's water supply.[34] To further protect the city from waste produced farther away from shore, over the next twenty years engineers built sewers on the south and north sides of the city that drained into the Chicago River, rather than the lake. This helped Chicago. Those living "newly" downriver did not appreciate the changes.[35]

Chicago wasn't the only city on Lake Michigan. Milwaukee, Wisconsin, to the north, and Calumet, Indiana, to the south, both had their own industrial production hubs and populations that utilized the lake's waters. Currents typically moved wastes dumped into Lake Michigan away from urban regions. Sometimes weather changed that. From 1903 to 1938 the Calumet region—which included Gary, East Chicago, Hammond, and Whiting, Indiana—saw a "phenomenal increase" in the number of industries located there; according to Arthur Gorman, an engineer working for the Department of Public Works of the City of Chicago, "It has been general practice in this area to discharge industrial wastes and domestic sewage without treatment directly into Lake Michigan or into waterways draining into the lake."[36] Samuel Greely, a consulting engineer at the Chicago firm Pearse, Greeley, and Hansen, estimated in 1921 that industrial sewage produced in the area amounted to "over 500,000,000 gallons per 24 hours."[37] More recent estimates highlight that between 1905 and 1910 the tons per year of phenol, cyanide, oils, and sulfuric acid being dumped into the Calumet Harbor (located on the southeast of Lake Michigan) approximately doubled, with levels increasing again in 1920, 1925, and 1930.[38] Lake Michigan's waters reflected the industries that used it. In short, the "terroir" of Chicago's water supply distinctly tasted and smelled of industry.

## MANAGING INDUSTRY'S TERROIR

The concept of terroir undergirds contemporary arguments for protecting traditional food production methods around the globe. Although the majority of efforts remain centered in European countries, people in South Africa, Peru, Mexico, and elsewhere have adopted the concept to promote traditional foods like rooibos, quinoa, and tequila.[39] Terroir-based promotion highlights the link between the taste of a food and the specific place a food is grown and produced; many promotion efforts also highlight specific labor practices designed to maximize the connection between taste and place. In contrast, as noted in the introduction, industrial terroir seeks to divorce taste from place. Industrial terroir comes about through technological, regulatory, and expert practices of *making unavailable* the sensory qualities that mark place-based uniqueness. Industrial terroir aims to minimize, rather than center, place-based uniqueness.

To improve the quality of raw water coming out of Lake Michigan into Chicago's municipal treatment facilities, the city's engineers had extended intake pipes farther out into the lake, constructed water cribs—structures designed to improve water quality by drawing water from deeper in the lake—and reversed the Chicago River to send wastes away from instead of toward the lake.[40] Despite the many engineering fixes put in place, Lake Michigan's raw and treated water continued to reflect the industries around the region. While waterways offered an easy disposal method for industries like Dow, the impacts of disposing waste into water proved more difficult to mitigate. Some effects were immediate, such as fish die-offs. Other effects were more subtle, showing up in bodies at a significant distance from the initial source. Of the many chemicals dumped into waterways, phenol produced in the process of distilling coal into coke and then discharged into lakes or rivers proved especially troublesome.

The terroir of Lake Michigan's raw water in the 1920s and 1930s was uneven.[41] Some communities regularly took in raw water that had perceptible traces of industrial action on a daily basis. Others did not. This difference was due to the fact that many of the industries shaping the terroir of Lake Michigan's raw water were located at the southeastern and southern edges of the lake, in large part because of the proximity to transportation routes. Waterworkers in Chicago understood this spatial distribution as limiting industry's impact on the water being drawn from Lake Michigan into Chicago proper—at least until winds, storms, or ice flows interfered with typical circulation patterns.[42] The tastes of raw water coming into a city could unexpectedly change when winds and lake currents moved phenol- and sewage-containing waters significant distances from their points of origin.

Waterworkers, engineers, and others working on providing municipal water were well aware that the "increased volumes of industrial wastes that are being poured into the rivers used as sources of public water supply" contributed to the need to eliminate or prevent taste and odor problems.[43] They lived in a time period characterized by rapid expansion of industrial pollution, as well as by an emerging conception of how much pollution the environment could assimilate.[44] Research building on the British concept of "self-purification" (the ability of the environment to remove harmful substances), conducted by Earle B. Phelps and H. W. Streeter on the Ohio River, argued that one could quantify how much pollution a body of water could assimilate.[45] By developing a system to determine the "capacity of a body of water to receive and dispose of sewage without the production of unsightly appearances or unpleasant odors," Phelps pointed out, researchers could determine the "true measure of a nuisance." They could also "determine in advance the final result of any anticipated pollution."[46] Streeter and Phelps's developed an equation that allowed researchers to estimate how much oxygen sewage dumped into water would consume. When water researchers graphed out the oxygen capacity of water challenged with increasing levels of pollution, they got a curve with an inflection point—a mathematical indication of when the oxygen in a stream can no longer break down pollutants. Streeter and Phelp's quantification, as Max Liboiron demonstrates, has come to represent the threshold where legal structures recognize pollution.[47] Defining pollution around environmental assimilative capacity, Liboiron argues, set in place "the foundation of the permission-to-pollute system" that defines how contemporary legal structures and regulatory policy manage waste.[48] "The logical extension of quantifying the threshold of pollution," Liboiron notes, "was to parcel out assimilative capacity—essentially the ability to waste, even the right to waste—to polluters."[49]

Lake Michigan did not always "assimilate" the wastes dumped into it. The "unsightly appearances or unpleasant odors" that emerged when winds or currents moved concentrated plumes of Lake Michigan's polluted southern waters northward defined the Lake's shifting terroir. The challenge for Chicago's waterworkers was to manage those perceptible, ingestible markers of industrial activity.

Making foods with industrial terroir is a practice of managing the ingestible markers of place so that they do not overly interfere with political aims or economic activity. Those markers may be perceptible. They may not. In the early decades of the twentieth century, especially after implementation of the Public Health Service's standards for water in 1914 (and subsequent revision in 1925), Chicago waterworkers primarily managed Lake Michigan's industrial terroir—with its mix of chemicals, sewage, and beyond—through chlorination. For example, Chicago's municipal waterworkers went from using approximately 673,000 pounds of chlorine to treat the system's water in 1923 to 1,250,890 pounds in 1924 in response to decreasing quality of the raw water supply.[50] To address public health concerns, the management of industrial terroir required managing chemical contaminants and microbial life. Uncertainty about the impact of increased chlorine levels on the "delicate tissues and organs of the human system" accompanied certainty about the destructive impact of free chlorine on metal pipes and machinery.[51]

Transforming the industrially inflected taste of Lake Michigan, its terroir, into a quality (read: safe) product did not always result in a product with palatable qualities. The chlorine added to kill bacteria, and thus make the water safe, could destroy some of the compounds that caused unpleasant flavors or odors (although many in the early days of chlorination found the odor of chlorine unpleasant). However, chlorine itself could combine with phenol to produce especially strong, unpleasant tastes in the water. As Baylis pointed out in a paper on taste and odor elimination, "taste tests are so unreliable in judging the efficiency of a treatment that it is difficult to state at the present time whether tastes other than those produced as a result of the addition to chlorine to water are actually reduced."[52]

Lake Michigan's uneven quality frustrated consumers. It frustrated waterworkers, too. Across the Chicago region, water customers on the city's South Side faced the brunt of pollution in Lake Michigan and as such were most exposed to industrial terroir. Sometimes that meant their water got "double chlorination," whereby initial treatment with ammonia chlorine proved inadequate and required a second, supplementary "minor chlorine treatment."[53] Increased chlorination degraded the aesthetic qualities of water flowing into homes, especially when chemicals like phenol were present. Such episodes resulted in medicinal-tasting water. While Chicago's overall water quality, as measured by typhoid cases, had improved through the building of increasingly distant intake tunnels and the reversal of the river, consumers on the South Side received a water supply "decidedly inferior to that in the north and central sections of the city."[54]

The "decidedly inferior" quality of South Side water impacted a diverse array of people. The 68th and Dunne Street water intake cribs sent water to neighborhoods where almost everyone rented as well as neighborhoods where almost everyone owned their homes. They provided water to Chicago's booming African American neighborhoods, most of which were on the South Side, and delivered water to foreign-born immigrants as well as longtime, unmarked (presumably white) inhabitants of the city.[55] The 68th and Dunne Street intake cribs even fed the plant delivering water to the homes of some of those tasked with managing the city's municipal water.[56] In short, a wide range of socioeconomic, ethnic, and racial groups received notably inferior water.

The turbidity, odor, and taste of water coming out of Chicago taps brought the perceptible indicators of industrial terroir—the chemicals found in raw water as well as the chemicals used to make that raw water safe—into the intimate walls of a range of households regardless of economic class, race, or ethnicity. Indeed, the *Chicago Daily Tribune's* food columnist, Mary Meade (a nom de plume), noted that "in cities such as Chicago, where chemicals are necessary additions to the water, to make it safe for health, frequently a very definite taste is thus imparted [to foods cooked] by the chemicals." Coffee, tea, and the pinnacle of mid-century homemaking, gelatin salads and desserts, Meade noted, were especially threatened by "the use of anything but clear spring water or other specially filtered water."[57] Industry, and water-workers' efforts to mitigate industry's discards, distinctly shaped the flavors found in Chicago's municipal water. Cooks shifted their behavior to minimize exposure to the flavors of interwar industrial terroir; no matter how safe the experts proclaimed the water, inhabitants such as cooks understood the waters' sensory characteristics as, at a minimum, threatening their culinary well-being.

Chicago's water producers did not understand their efforts to mitigate the effects of industry on the quality of Lake Michigan's water as creating industrial terroir. However, the chlorine used to make water potable shortened the life span of the machinery used to treat water and increased operation costs. The unpleasant tastes and odors resulted in public frustration. As waterworkers sought to minimize the impact of industrial activity around the south side of Lake Michigan on the raw water coming into their treatment plants and thus on their individual and community reputations, they worked toward a future in which their control over the tastes caused by industry and treatment would fade away. One of their first steps was to pressure city officials to help obtain funding for building a filtration plant on the south side.

## FILTRATION DREAMS

The idea of filtering water to improve its aesthetic and bacteriological qualities was not new—filtration is one of the earliest and most basic water purification techniques known. Indeed, the phenol wastes plaguing Chicago, especially Chicago's South Side, were a smelly reminder of a larger, decades-long political battle over the city's municipal water treatment infrastructure. Since the early 1900s, future-looking commissioners and engineers had lobbied the city council for funds to support development and building of filtration plants to help address the ongoing challenges of managing Lake Michigan's industrial terroir.

Central to this challenge was Chicago's system of charging for water. In the mid-1920s, when the commissioner of public works, Colonel A. A. Sprague, requested a report from the city engineer on Chicago's water quality, water customers paid a flat rate for their water. "The only real obstacle in the way of filtration, from an economic standpoint, is the city's excessive per capita water consumption, which could be corrected by metering. On any other basis the costs would be very high and necessitate a substantial increase in water rates" the commissioner opined after the late 1932 and early 1933 repeat of phenol contamination in the city's water.[58] Filtration increased the cost of producing water. However, the flat rate disincentivized consumers from changing their behaviors. As the cost of treating water increased as demand increased, waterworkers reiterated the need for a shift in how water usage was billed. For filtering to work, waterworkers said, users needed to stop their practice of running faucets continually to ensure the coolest, freshest water; fix leaky plumbing; and generally find a fiscal reason to more carefully use their water.[59] While it is unclear to what extent each practice contributed to loss of treated water, analysis of usage by metered versus unmetered consumers demonstrated a marked difference.[60] The image of water flowing unchecked from treatment facility, through city pipes, out taps, and into the sewage stream certainly carried some rhetorical weight, even as some consumers, and politicians, resisted such efforts.[61]

Just as the perceptible aspects of Lake Michigan's industrial terroir helped spur reversal of the Chicago River, distasteful tastes and odors offered a central talking point in the effort to implement metering and build a permanent filtration plant. Engineers pointed to the intermittent, headline-grabbing moments when medicinal phenol tastes and odors or other contaminants moved the quality of water delivered from acceptable to mildly unpleasant or offensive. City engineers—responding to and drawing on complaints of funny-tasting water—successfully lobbied Chicago's city council in 1926 to begin experimental research on filtration; the council allocated $50,000 to support the effort.[62] By 1927 the experimental filtration plant was operational.

To spearhead the experimental work, the city needed expertise it currently lacked.[63] Loran Gayton, the engineer of Chicago's Water Works Design division, hired John Baylis. Trained at Mississippi State College in railroad engineering and construction engineering for water and sewage plants, Baylis was known for his work on measuring turbidity (cloudiness) and improving the quality of Baltimore's municipal water supply—a difficult task given the polluted surface sources supplying the city.[64] Baylis estimated soon after his hiring that Chicago's water was among "the most turbid water being supplied to any large city in [the United States]," a problem, he dryly noted, that left "clarity and taste not all that is desired."[65]

Baylis's hiring marked an inflection point in the management of the industrial terroir of Chicago's raw water supply: over the subsequent years, he and his colleagues would test a range of methods to improve water quality. "The tentative program" Baylis reported in 1927, "involves such problems as the effect of microorganisms on coagulation, removal of tastes and odors, removal of microorganisms, various kinds of coagulants, optimum chemical conditions for coagulation, the value of mixing in forming a flocculation, types of mixing basins," and many other issues.[66] The experimental program's usage of superchlorination, followed by dechlorination, and ensuing treatment with lime "makes city's water taste like a spring" reported the *Chicago Daily Tribune* in 1930.[67] Baylis's experimental approach offered a sensory argument for the ability of waterworks to manage industrial terroir.

To increase public support for the proposed filtration plant, the city made filtered water freely available to anyone willing to drive to the experimental filtration plant. One report indicated that "from 600 to 1000 automobiles a day visit this plant and carry away with them an average of from five to seven gallons of water."[68] On days when phenol contamination caused medicinal flavor, the number of cars could go as high as 12,000 per day. "The streets are crowded for blocks," a news report noted, pointing out that some people came as far as ten miles to obtain aesthetically pleasing (or at least inoffensive) water.[69] Waterworks officials hoped that this freely available water would act as a gustatory rhetoric capable of overcoming public opposition to metering (seen as critical to filtration) and the costly price tag of the plant: one engineering consultant estimated the cost at approximately $60,000,000 in 1934![70]

Due to the anticipated cost of filtration, the proposals argued for first building a full-sized plant on the South Side. Additional plants would then be constructed as money allowed in other areas around the city; plans submitted in 1925 also suggested plants at Wilson Avenue, Chicago Avenue, and 12th Street.[71] Recognizing that the proposed filtration plants would require significant amounts of land, initial proposals suggested making the plants "esthetically attractive" and multipurpose by including recreation facilities.[72] Proponents' dreams of activating an aesthetic rhetoric where filtration plants looked beautiful, offered recreational possibilities, and made the industrial terroir of Lake Michigan's water into something inhabitants valued withered in the face of Chicago's city council. Newspaper accounts of city council meetings from the 1920s through the mid-1940s read like a soap opera: promises made, withdrawn, threats offered (by the federal government), misunderstandings resulting in uproar, and more. One opinion letter writer proposed abandoning the idea of filtration and instead turning "efforts to making [contaminating factories] cease contamination of our own source of supply" rather than wasting "millions and millions to clean up [the polluters'] mess."[73] Construction on the South Side for what was then "the world's largest filtration plant" finally began in 1938, despite earlier citizen opposition from neighborhood associations. The entrance of the United States into World War II slowed completion significantly, an unfortunate gustatory reality for South Side residents as wartime industrial activity increased in the south Chicago region and northern Indiana. When finally completed in 1947, the South Side filtration plant offered residents improved beach access adjacent to the filtration plant. Unfortunately, the earlier proposed offering of recreational facilities that directly combined with water treatment had largely disappeared.[74]

Waterworkers' conversations with each other in the interwar period focused on treatment and removal rather than on agitating for regulation to undo the actions that created the industrial terroir of Chicago's water. Baylis, for example, embraced a pragmatic view of manufacturing industries: "Most industries have certain waste products that must be disposed of in some manner," he stated to the Minnesota Section meeting of the AWWA in October 1931, "and sometimes it is possible for the industry to exist only where its waste products can be disposed of cheaply. . . . [I]t is impossible to eliminate all industrial pollution of objectionable character."[75] Baylis pointed out that in addition to coke oven plants and oil refineries, canneries, textile mills, milk product plants, paper mills, glue factories, beet sugar mills, meatpacking houses, steel mills, tanneries, and dye and alcohol manufacturing plants all had waste products that could cause disagreeable tastes. However, instead of suggesting changes to industrial waste disposal, a step Baylis saw as potentially leading to the loss of employment as well as regional power, he suggested "a spirit of cooperation between inhabitants and industry." Key to this spirit of cooperation? "Water works officials should be alert to developments in the treatment of polluted water and be ready to meet increased pollution should it come as a result of increase in industries and in population."[76] Baylis highlighted to his audience how different treatments used to remove dangerous bacteria could also be used to make water taste and smell better. He aimed to mobilize waterworkers' ability to catch the perceptible markers of the places water came from, to reliably test that water, and to treat it to remove as many markers of place as possible. For Baylis, the spirit of cooperation between inhabitants and industry depended on the ability of waterworkers to use engineering and chemistry to shift the physiological point at which people noticed and objected to their water, while continuing to rely on waterworkers' expertise in protecting public health.[77]

## QUANTIFYING INFLECTIONS

Managing the industrial terroir of Chicago water characterized Baylis's everyday research activities. From 1927 onward, he and his colleagues focused on experimenting with different treatment techniques to improve the aesthetic qualities of the water they produced. At the time, approaches to improving water quality varied widely among different municipalities. For example, reports and discussion at the AWWA's Toronto convention in 1929, and as part of a panel presentation before the Water Purification Division at the St. Louis convention in 1930, indicate that some found activated carbon useful in their taste and odor removal efforts. In contrast others, like the water supply engineer for The DuPont Company, found carbon filters unable to remove "bad" and "very bad" chlorophenol tastes from the company's drinking water.[78]

The experimental filtration plant allowed Baylis and his colleagues to make water that subjectively demonstrated the effectiveness of various treatment techniques. Yet Baylis and his wider swath of professional colleagues lacked the ability to evaluate whether their experimental approaches could effectively travel to different locations. At the 1931 Minnesota Section meeting of the AWWA, Baylis pointed out that an operator at one location could report success in removing "decaying vegetation, earthy, fishy, grassy, moldy, swampy" tastes or odors. Another, located elsewhere, might face similar taste or odor challenges but report that the same treatment others had used was ineffective. Lack of an "accurate means of determining when the water is free from taste, or of determining the intensity of the taste," posed a challenge to waterworkers trying to remove unwanted sensory cues.[79]

Waterworkers could talk about taste. Yet they found that the then current accepted method—codified in the *Standard Methods for Examination of Water and Sewage*—failed to help them compare how well a treatment worked. First introduced in 1899, the method suggested that waterworkers grade odors on a scale of 1–5. A sample with an odor of 1 was "very faint," an odor undetectable by the "average consumer" but perceived by a waterworker in the laboratory. In contrast, a sample graded as a 5 was so intense that no one would willingly choose to drink the water.[80] Baylis pointed out in a presentation to fellow waterworkers in 1931 that even when multiple experienced operators tested the same sample of water, they struggled to come up with the same ratings.[81]

Indeed, the *Journal* of the AWWA is filled with reports of experimental efforts to eliminate or prevent off-tastes and odors.[82] Some recounted experiments in aeration, others various types of filtration or application of differing levels of chlorine. In the latter half of the 1920s, additional treatments such as the addition of ammonia before treatment and application of activated carbon—an innovation Baylis played a key role in testing and implementing—emerged. Operators publicly complained to each other in print and at section meetings about the difficulties of evaluating various experimental methods used in one place for efficacy and for economy in another place. After all, operators at different water treatment plants faced unique treatment challenges. What was "economical" for any single plant depended on a variety of factors such as the weather, operator skill, location, what local people liked, operating budget, and most pressingly, the nature of whatever was causing the problem. Even as treatment processes for making water safe to drink became increasingly standardized and capable of circulating in handbooks and journal articles, the ability to evaluate funny tastes or smells in water remained stubbornly linked to whether individual operators could reliably use their senses to evaluate the efficacy of any given treatment approach in managing the industrial terroir of their water supply.

Baylis was not alone in his frustration with the existing method for measuring reduction of unwanted tastes and smells. Charles H. Spaulding, of the municipal water plant in Springfield, Illinois, publicly voiced his skepticism about the hopeful rhetoric accompanying reports of new treatment methods. Like Baylis, Spaulding worked with surface waters contaminated by industrial production and human waste. Like Baylis, Spaulding was interested in and actively researching filtration. "The trouble has always been," Spaulding commented, "that we were not able to measure odor, or taste, quantitatively."[83] This made it difficult to identify to what extent mitigation efforts worked, even as accounts accumulated of increasingly successful removal of off-tastes and odors via various treatments. Experimentation without quantification, Spaulding complained, was nearly useless.[84]

At the core of Spaulding's complaint lay a long-standing concern among scientists about the subjectivity implicit in sensory experiences of smelling and tasting. The already existing systems used to measure taste and smell, introduced at the turn of the century, relied on waterworkers developing some level of expert connoisseurship about the water they worked with.[85] Unfortunately for waterworkers interested in a scientific way of communicating about their experience, connoisseurship works best when it relies on individual, well-trained experts. It depends on individual experiences accumulated through repeated encounters with an object. Indeed, connoisseurship was, and is, a practice central to the construction of the concept of terroir in France, in part through the work of defining one product's qualities in contradistinction to those of another product. For example, as Thomas Parker points out in his history of the idea of terroir, when a wine connoisseur describes a wine from one region as tasting of cherries, and another from a neighboring region as "gamey," they are "replacing vaguely perceived sensorial qualities in the taster's mind" with "solid, discernable, contrasting identities: 'cherries' vs 'game.'" By using words like *cherry* or *gamey* and the images that accompany them, Parker suggests, connoisseurs increase for themselves and others the differences between products from different regions.[86]

Connoisseurship, so central to the concept of terroir, relies on a person's ability to describe personal experience. However, connoisseurship does not easily circulate: it depends on individual levels of expertise around tasting and naming differences. It's not a surprise that Spaulding saw the method from *Standard Methods* as simply one worker's "expression of opinion." Taking direct aim at the approach currently in use, Spaulding called for a "yardstick," a means of determining odor that was not the "opinion of the observer based on a momentary impression which cannot be recorded and referred to for comparison."[87] As a subjective and intersubjective practice, water evaluation using a connoisseurship approach failed to create "objective" knowledge devoid of the observer—the gold standard increasingly sought by scientists.[88] Spaulding proposed moving away from a connoisseurship model, suggesting that waterworkers instead adopt a method, the TON, for determining the threshold level at which an odor was just barely detectable. "We [waterworkers] cannot as yet determine quantitatively many of the actual compounds causing odor," Spaulding pointed out. Waterworkers like Spaulding saw efforts to manage water quality based in connoisseurship-like approaches as incapable of helping solve the pressing taste and odor issues undermining their efforts to build systems that could balance between the different needs of their communities.[89]

## FINDING THRESHOLDS

Although not explicitly stated, Spaulding's threshold approach to measuring the presence of off-tastes and odors drew from the growing field of psychophysics. Psychophysics emerged in Germany in the mid-1800s through the work of German researcher Ernst Heinrich Weber and his student, Gustav Fechner. Interested in the relationship between sensation, stimulus, and aesthetic judgment, Fechner drew on Weber's work examining how humans perceive touch to develop a mathematical law for measuring just noticeable differences (JNDs) in sensations such as pressure, strain, noise, and brightness. Using this law, Fechner set off on a lifelong pursuit of attempting to experimentally measure aesthetics.[90]

Fechner posited that aesthetics could be understood in "two fundamentally different ways": either as aesthetics "from above" or as aesthetics "from below."[91] Aesthetics from above, Fechner argued, worked deductively, allowing a researcher to deduce individual experiences from general ones. In contrast, aesthetics from below—the approach Fechner proposed building an entire research program on—worked inductively, drawing on individual terms and experiences to generalize to a population-level experience.[92] Fechner saw the deductive approach as making it "difficult to clarify the causes of pleasure and displeasure in particular cases" in its focus on generalities.[93] In short, the focus of his aesthetics from below research program was "to detect sources of inherent delight and appeal [of things] as well as the laws of their interplay." This could, he hoped, help "identify a general source, a general law of how pleasure and displeasure come to pass."[94]

Fechner named his research field *psychophysics*, pointing out through the term the ability to measure physical data and in the process obtain information about psychological experience. The idea that individual sensory thresholds could be identified and aggregated together undergirded Fechner's early efforts. It shaped his subsequent work seeking to understand aesthetic preferences. These thresholds were the point when a change in sensation is registered—Fechner posited that very small changes could result in detection of differences for very small stimuli, but that a much larger change in magnitude of a signal was necessary for detection of difference of a large stimulus. In other words, the weight of a fly is noticeably different than no weight at all, but adding a single book to a stack of one hundred books may not be perceived at all. Despite his mentor's interest in understanding sense discrimination for touch, temperature, weight, vision, smell, and pitch, the law Fechner derived from Weber's experimental work—a law that helped identify thresholds of perception—did not easily extend to all sensations.[95] Some stimuli (e.g., brightness intensity) proved easier than others (e.g., tone) to modulate at very minute levels. Repeated exposure to other stimuli resulted in physiological adaptation, with one's ability to detect the stimulus decreasing over time. This meant that while physical measurement of a stimulus would show an increase or decrease, a person's physiological reaction would show no change in awareness. As such, although multiple researchers had built on Fechner's work, it wasn't until the early twentieth century that scientists began investigating whether the law proposed by Fechner, Weber's law, could be applied to smell.

A young doctoral student at Cornell, Eleanor Acheson McCulloch Gamble, first took on what one reviewer of her research termed the "painstaking" work in a "somewhat unattractive field" of testing whether Weber's law was applicable to smell.[96] Gamble took advantage of a recently introduced instrument, the olfactometer, to isolate smelly compounds from the larger environment for direct delivery to the nose.[97] Gamble's account of her experimental efforts is a rich tapestry of the difficulties associated with scientifically measuring smell and of the general messiness of true experimental effort. Determining whether Weber's law applied to smell proved especially challenging: Smell was one of the stimuli people adapt to; In addition, many materials absorb smells. This meant that Gamble and her team had to carefully and repeatedly clean the olfactometers they used between each experiment. Furthermore, even slight shifts in how quickly and deeply one breathed had an impact on Gamble's experimental findings.

Gamble was entirely aware of the challenges facing her. "The peculiarly unsatisfactory character of the determinations of olfactometry (the branch of psychophysics concerned with measuring smell) and odorimetry (a 'side-issue' of olfactometry, concerned with the measurement of the intensity of smell-stimuli considered as objectively as possible) is due chiefly to the fact that olfactory qualities, unlike visual and auditory, are not demarcated," Gamble pointed out in the introduction to her dissertation. Technical instruments could measure decibels and lumens, while humans, with or without olfactometers to help direct odorants to the nose, remained the central, fickle instrument capable of measuring olfaction.

At the heart of Gamble's work was the effort to identify a "stimulus limen," the threshold at which odor perception (for different odors) occurred, and the JND, the threshold at which a person could identify a difference 50 percent of the time. Her work, despite challenges, demonstrated that threshold data *could* be obtained for different odors. More critically, it suggested that Weber's law could be applied to smell. This set in place a basis for others to find a way to put into action the insights that Weber's law offered to those working on odors—to extend, if you will, Fechner's hope of identifying how pleasure and displeasure come to pass into the practical realm of managing the relationship between water supply, industrial and municipal discharge, natural events, and end consumers. In other words, identifying the inflection point at which odor perception occurred also offered the hope of increasing waterworkers' ability to act on industrial terroir.

## FROM THRESHOLD TO PROACTIVE ACTION

"[I] claim very little originality in applying the dilution method to water," Spaulding wrote. "After adopting the idea, I found it was well recognized in chemistry and that the International Critical Tables contain lists of odor values assigned to various chemical compounds based on threshold value."[98] While Spaulding's inspiration appears to have come from Hendrik Zwaardemaker's work rather than Gamble's more explicitly psychophysical approach, it nonetheless bears the hallmarks of a psychophysical understanding of the world. In adapting a threshold approach, Spaulding adopted the belief embraced by psychophysicists that one can learn about the unknown through measuring the known: in this case, one could learn about the unknown properties of a taste or odor by identifying the known point at which one could first distinguish odor-free water from water containing the unknown, odor-causing compound(s).

For waterworkers, the questions of absolute stimulus thresholds and difference thresholds were questions of immediate and future action.

Immediate action came into play as workers analyzed the raw water coming into their facilities and determined how to treat it. Future action came into play as waterworkers navigated the tension between scientific, technological, and fiscal limits. After all, no amount of know-how could make up for city councils not approving increased budgets or rate changes. All of these modes of action shaped the relationship between a waterworks and the metropolitan region it served.

Recognizing the link between taste, odor, and the larger economic and political context of water provision, the AWWA formed the Committee on Control of Taste and Odor to report to the community on the state of the art. Its opinion was dire: "The entire subject of tastes and odors is so much a matter of opinion that merely summarizing the reported experiences of plant operators fails to give more than a resume of the honest opinions of these workers, and brings home the fact that these honest opinions conflict in many cases."[99] The problem, committee members argued, was due in part to the "limitations of the present method of odor determination in Standard Methods, which relies too much upon the personal equation as to quality and intensity."[100] The committee's report highlights the core challenge of researchers in the twentieth century interested in how bodies interact with the environment. Just as Gamble pointed out in her own research, the sensory experience of one subject does not easily result in reproducible experiences for another. For waterworkers, this challenge meant that although individual operators at various plants could authoritatively report on the success of a method, those insights failed to help compare different treatment methods. To address this concern, committee members suggested that the AWWA adopt "the Spaulding method for determination of the taste threshold point—the maximum dilution with tasteless water which will give taste." Doing so, they pointed out, "would be a step in the right direction."[101]

The right direction the committee highlighted was the direction of identifying sensory thresholds. Spaulding's 1931 method, although sparse in details, called for making up a series of dilutions of an unknown sample with odor-free water. Odor-free water, one commenter would later point out, was an accepted fiction: the preparation method of boiling until approximately 10 percent of the volume had evaporated, and then cooling, could not truly deliver water without odor—just water with *less* odor.[102] Examiners would then compare samples of interest with the odor-free sample. These initial dilutions, all made to a total volume of 100 mL, allowed one to estimate the range of perception, with additional samples being prepared to more closely isolate the dilution point at which the unknown sample was indistinguishable from the known sample. "We have not thus far noted any marked discrepancy in the results obtained by this method in the hands of trained, conscientious observers," Spaulding reported.[103] Shortly after the committee's recommendation that the AWWA adopt Spaulding's method, Gerstein and Baylis's colleague, Oscar Gullans, proposed to the Illinois Section meeting that a modification of Spaulding's method would be even more successful at allowing detection of odors that up until then had frequently been missed by waterworkers but detected by consumers. Rather than relying on an initial set of dilutions to try to get the range as Spaulding had suggested, Gullans advised drawing on the previously established method to identify the intensity of the odor and then using that to "make an estimate of the probable 'threshold points.'"[104] In other words, Gullans argued that he had an even better route to identifying the just noticeable.

Spaulding's method and Baylis's and Gullans's later modifications were quickly brought into the official methodological approach: the seventh edition of *Standard Methods* (1933) included it under "Nonstandard Methods"; three years later it appeared as a standard method in the eighth edition (1936). By 1937 Gullans was regularly using the "improved" threshold test to evaluate different odor elimination treatments. While cautioning that the results he obtained were simply "correct for the particular water used in the test at the time and in the manner conducted," Gullans was confident that if operators used similar equipment and followed the same procedure, "it is possible for any water works operator to determine the most effective treatment to eliminate objectionable odors." Elimination for Gullans was not complete removal of odors. Rather, he understood elimination as the ability to reduce the odor "to a threshold of 2, at a temperature of 70°C"—the point at which water workers understood most odors to fall below the average consumer sensory threshold.[105] By 1938 the AWWA Subcommittee on Specifications and Tests for Powdered Activated Carbons codified usage of TONs to determine the amount of powdered activated carbon necessary for effectively treating water to control odor. "The Threshold Odor Test is capable of practical application in the control of tastes and odors, and in the evaluation of competitive carbons," it noted, situating the test within a small cadre of approved evaluation methods.[106] Using the test would not only supplement the commonly used phenol test, the committee pointed out, it would also allow operates to verify their approaches, providing "predictable performance in actual taste and odor removal from natural supplies."[107]

Reports from the late 1930s and early 1940s indicate that adoption of TON depended in part on operator skill set, municipal size, quality of source water, and research aims. The plant at Hammond, Indiana, described the threshold odor test as "very important because of the pollution of the source of the water supply, the southern end of Lake Michigan" and used it to determine how much carbon to add to raw water.[108] In contrast, tests of granular activated carbon performed by the Bay City, Michigan, superintendent of filtration included no reference to TON, instead relying on reduction of measured phenol presence as an indicator of reduced odor.[109] However, the Bay City example seems an exception rather than the rule given its reliance on a secondary measure.[110] For example, a survey examining taste and odor control on Lake Michigan sent out in 1939 to water treatment plants along the western shore of the lake relied on odor numbers to make its case about how water in the lake varied across locations. Reported TONs demonstrated the effectiveness of various treatments at removing unwanted tastes and odors caused by microorganisms or industrial waste. Many of the plants responding were using activated carbon to address off-tastes and odors in raw water, with efficacy demonstrated via pre- and posttreatment odor numbers. For the report's author, the chief chemist at Milwaukee Water Works, the combined findings of the survey, along with his own plant's research examining the temperature and odor threshold of raw water, proved that "if rigid control of the treatment [enabled through threshold odor testing] is properly maintained, a palatable water can be produced."[111] The threshold odor test helped waterworkers in geographically disparate locations compare the efficacy of different treatment methods in reducing tastes and smells. By combining the threshold test with other forms of experimentation such as the testing of filtration techniques, usage of activated carbon, prechlorination, and so on, workers in Chicago demonstrated they could act to remove offensive sensory cues before water ever left the plant; they became even more active in shaping the industrial terroir of Chicago's water.

## INCOMPLETE INNOVATIONS

The historical actors highlighted in this chapter did not articulate their efforts under the framing of industrial terroir. Rather, industrial terroir emerged through their responses to the impact of industrial action on the natural characteristics of Lake Michigan. The sheer size of the lake, as well as its geographic location, facilitated industrial growth. Yet the lake's natural flows didn't allow the discards from industrial activities or human life to completely disappear. Instead, those discards created an ever-shifting set of characteristics of the raw water feeding Chicago and other cities along the lake. During the interwar period Baylis and his colleagues developed and tested a significant number of innovations in transforming the qualities of raw water flowing through treatment facilities to residents of Chicago's South Side. These experiments ranged from chemical treatments such as superchlorination to physical treatments like activated carbon filtration, all validated by TON measurements. The successful diffusion of those treatment innovations beyond the experimental filtration plant depended in part on waterworkers' ability to *measure* the effectiveness of treatments. By drawing on theories about physiological thresholds—theories themselves rooted in an understanding of sensory experience as aesthetic—Baylis and his colleagues were able to mobilize TONs to show the larger sanitary engineering community that they could remake raw water's perceptible qualities into increasingly imperceptible qualities. In doing so, they put in place the possibility for a future in which unwanted, unfamiliar, or disagreeable tastes and odors could be completely mitigated, a process that would shift who could be aware of the quality and qualities of the raw water feeding the city.

The innovations Baylis and his colleagues developed incompletely transformed the perceptible qualities of water from Lake Michigan. Waterworkers' active management of water quality and water's qualities set in place the steps for shifting the industrial terroir of water delivered to the South Side. Those shifts came about not because of a change in what industrial producers or inhabitants put into the lake (although such changes would increasingly happen in the late 1940s and beyond). Those shifts came about because of changes in the making of municipal water. Rather than reflecting taste of place, industrial terroir reflects the ways of managing taste of place so that place itself can fade away.

By the early 1940s, Baylis, Gullans, Spaulding, and others saw in their work a near, attainable future in which the municipal water they made delighted rather than offended consumers, as long as waterworkers could effectively standardize the procedures enacted *and* the individuals making measurements.[112] Subsequent decades would demonstrate the deep challenges of such a task. As the next chapter explores, working toward fulfilling the future anticipated by Baylis, Gullans, Spaulding, and others required more than changing treatment approaches or building better filtration facilities. The partial realization of that anticipated future depended on a significant shift that occurred from the 1950s to the 1980s in how scientists and engineers acquired knowledge about what tastes and smells were and how to manage them.

# 2

# MAKING FLAVOR MOLECULAR

IN AUGUST 1979, the Metropolitan Water District (MWD) of California experienced its worst water taste and odor crisis since the 1940s.[1] As reports accumulated of musty, earthy flavors coming out of taps around the region, MWD's engineers set about looking for the source of the odor-causing molecules. "We did some research and didn't find much about it but had some idea that maybe it had to do with algae production, but we didn't see anything really changing in the water column in the planktonic algae," recalled Michael McGuire, the engineer in charge of the Water Quality Laboratory at MWD at the time of the crisis; "We had no idea what was going on."[2] Waterworkers took samples from all over the system in an attempt to understand what was happening. Their efforts floundered. The earthy, musty odors continued appearing with no apparent pattern throughout the distribution system.

Lack of evidence left waterworkers unable to confirm their suspicions. The MWD responded by sending divers, briefed by microbiologists about what they might look for, to Lake Mathews, California, to search for the source of the problem.[3] Unlike Lake Michigan, with its close proximity to Chicago-area drinkers, Lake Mathews lies far from the mouths and noses of most Southern California water consumers. Built in the late 1930s and filled in the early 1940s to store water from the Colorado River, Lake Mathews is a key reservoir for Southern California's water supply.[4] By the 1950s, a combination of drought and agricultural runoff had negatively impacted the quality of water stored in the newly created lake.[5] Waterworkers suspected that despite the distance, Lake Matthews might be where the odd tastes and odors originated from. "I'll never forget," recollected McGuire. "It was a nice sunny day and we were standing on the shoreline and the divers were in the water, and one of them came up with this kind of green goo on his hand and he said 'Is this what you're looking for? Is this what it smells like?' And we took one whiff of it and of course it was just dramatically a strong, earthy musty odor."[6] The diver held samples of an attached blue-green algae found on the bottom of the reservoir. As far as the researchers could tell, up until that moment no one had identified taste and odor problems caused by this type of algae. Green goo in a diver's hands and the nose of a frustrated scientist combined to connect the smelly problem plaguing MWD's member agencies with a blue-green algae.

When the muddy, earthy, musty odors started flowing through the pipelines and treatment plants of Southern California, those responsible for treating water found it difficult to detect what was causing off-flavors. The tools workers at the MWD had both differed from, and were startlingly similar to, the tools used by John Baylis, Oscar Gullans, Charles Spaulding, and others in the interwar period. By 1979, McGuire and his colleagues found evaluating water quality easier than Baylis and his colleagues had, thanks to improvements in instrumental detection. At the same time, the TON, which the profession incorporated into *Standard*

*Methods for the Examination of Waters and Sewage* as an official method in 1946, had been adopted and remained the main method for measuring taste and odor problems as well as evaluating the effectiveness of treatment.[7] Yet by 1971—shortly before passage of significant amendments to the Federal Water Pollution Control Act—the editors of *Standard Methods* had completely dropped all odor descriptors from the manual.[8] "A satisfactory system for characterizing odors has not been developed despite efforts over more than a century," the editors noted, categorizing continued efforts to describe tastes and odors as fruitless.[9]

The molecular culprits responsible for the musty, earthy odors didn't show up in tests at the place of production. Rather, they only appeared via an aggregate of consumer complaints. The inexplicable randomness of the complaints further confused the picture: only some of the member agencies of the MWD reported problems, despite the fact that all members received water from MWD.[10] Waterworkers lacked the necessary data to solve their problem even as the complaints made it apparent that without a new way of understanding off-tastes and smells, water suppliers would inevitably find themselves submitting "endless samples" for analysis without gaining any further insight into the source of the problems.[11]

Because mouths or noses can detect some smelly compounds at levels well below instrumental sensitivity, human sensory systems became the most likely source for detecting the presence of unwanted tastes and odors. Unfortunately, like the analytical instruments that had failed to offer workers at MWD clues to the odorants plaguing their system, the sensory-based TON test also proved inadequate. The hopeful rhetoric put forth by early waterworkers like Baylis, Chicago's engineer in charge of monitoring quality and an early proponent of the TON method, clashed with the reality that TONs are unique to each taster. This meant that the numbers generated only identified the *range* of detection thresholds; the method provided waterworkers with information that still resisted being distilled down into a single number that could lead to definite action. To further complicate the situation, TONs obtained using pure standards in the controlled conditions of the laboratory did not work well in the field: odors, when mixed together, can interact in unexpected ways. They might produce odors greater than the sum of the individual parts (synergism) or might cancel each other out (antagonism).[12] In addition, the most intense odor identified during TON analysis (and thus the odor that the number was associated with) might not actually be the odor causing the problem. In contrast, an odor molecule at a much lower concentration could result in intense consumer dislike.[13] At best, the TON was clumsily suited to the challenges of mid- to late twentieth-century pollution. This meant that tastes and odors found in water decreased in scientific importance at the same time that industrial pollution increasingly complicated efforts to manage potability and palatability.

"The problem with [TON] is it doesn't tell you anything about what's causing the odor, which we know are chemicals, and it doesn't give you any idea how to treat the odor," chemist Mel Suffet remembered when asked about odor detection methods.[14] This was precisely the problem facing the MWD: "We didn't know anything, what was going on," McGuire recalled. "We had no clue, the only thing available was the Threshold Odor Number, which was a really lousy test for taste and odor."[15] The failure of already existing sensorial and microbiological monitoring methods to detect problems, coupled with the massive levels of consumer complaints, called for new approaches.

Although McGuire was aware of the human limits of detection, he recognized that human mouths and noses offered an untapped sensitivity for detecting and identifying off-flavors. After all, it was customers annoyed by the sensorial information found in their drinking water who were flooding member agencies throughout Southern California with complaints. McGuire and his colleagues at the MWD took an unusual step: they crossed disciplinary lines, turning to consultants from the food and consumer industry to help them figure out how to identify and erase pesky olfactory or gustatory problems before water left the treatment plant.[16] The moment when the diver held up some green goo and muck to McGuire marks an inflection point in thinking about, analyzing, and developing plans for treating water. Inflection points happen not only when professional groups adopt new approaches such as the TON. They also happen when the knowledge and approaches from multiple industries come together in a new mixture. The coming together of approaches from disparate fields opened up new pathways for thinking of water more like a food, which centered on adopting a different understanding of taste and smell in molecular terms.

## EPISTEMIC CULTURES

The MWD, specifically, and the municipal water industry writ large, were not alone in facing unwanted flavors and odors. Food producers faced similar difficulties. The move to industrial-scale production, with its constant recipe reformulation and processing methods aimed at extending shelf-life, had come at the cost of many flavor components. Traditional methods of responding to unwanted tastes or smells failed when producing food at scale: throwing out a quart of burned sauce is one thing, disposing of hundreds of thousands of gallons of off-flavored sauce quite another. Just as scaling up recipes called for new techniques and levels of precision, the problem of off-tastes and odors invited those working on food in private industrial spaces and public research facilities to collaborate in the creation of a new culture of knowledge-making. This culture centered on an understanding of flavor as molecular, as coming into being only when a flavorful molecule binds to a receptor in the ortho-nasal passage.

The idea that flavor only exists when certain kinds of molecules come into contact with the receptors in a nose or mouth may be familiar for those of us living in the twenty-first century. But it wasn't always this way. This is not to say that chemists and perfumers don't have a long history of linking certain odors with certain molecules; they do. However, thinking of flavor—the combined experience of tasting and smelling something—as a molecular phenomenon is relatively recent, a development deeply rooted in what historian David Baird calls the "instrumental revolution in analytical chemistry" that occurred during the twentieth century.[17]

To understand the role instruments played in shifting the granularity of data available to waterworkers during the late twentieth century requires understanding how innovations during the mid-twentieth century helped create a new epistemic culture of flavor. Karin Knorr Cetina defines epistemic cultures as "amalgams of arrangements and mechanisms—bonded through affinity, necessity, and historical coincidence—which in a given field, make up how we know what we know."[18] Epistemic cultures come into being when a particular combination of scientific tools, methods, regulations, and ways of asking questions become *the* accepted approach for investigating the world. Once an epistemic culture comes into being, it can be hard to imagine that anyone once thought otherwise.

As scientists in the food and flavor industry and associated fields worked to understand tastes and smells, they became increasingly interested in the molecular identity underlying sensory experience. Identifying molecules, however, remained tricky until the mid-twentieth century. The work of making sense of sensing changed with the invention of new instruments such as the gas chromatograph and mass spectrometer. These instruments helped solidify a new understanding of taste and smell as molecular.[19] Made molecular, flavor became granularized, a data point increasingly amenable to technical manipulation.

## DESCRIBING FLAVOR

The work of making flavor molecular occurred in multiple places. It brought together multiple disciplines. The most obvious starting point, however, is the joint work spearheaded by the US Army in the 1950s in response to soldiers' resistance to rations during World War II.[20] Nutritionists had considered the foods developed for troops during World War II nutritionally complete: each meal claimed to provide the necessary carbohydrates, proteins, fats, vitamins, and minerals for ideally functioning bodies. Yet soldiers disliked many of the meals.

"Ours is a problem of measurement," Dean Foster said in his opening statement at a two-day conference jointly sponsored by the two army divisions in October 1953. Foster and the other academics, industry researchers, consultants, and members of the Food and Container Institute for the Armed Forces (established at the end of World War II) had gathered together to address the fact that all their measurements of "acceptability and excellence of the quality of foods and beverages" had failed to result in soldiers liking their rations.[21] "How can we advance the research program which is so necessary, and how can we supply the Armed Forces with rations of high acceptability?" Foster asked. He then turned his rhetoric to the general population, asking whether the group was ready to provide "the civilian public with food products of known excellence of quality?" He pointed out that "the Military, along with the food manufacturer, distributor, and seller, have come to recognize the importance of measuring food acceptance. They know that human welfare and purchasing behavior are affected directly by the acceptability of the available products."[22] Throughout the two-day symposium, discussions centered on measuring and articulating taste and smell. The discussions addressed the assumption that if one could reliably communicate about tasting, one could also begin to solve the gustatory puzzle of how to measure consumer acceptance of food.

Contemporary sensory science textbooks report that the field's growth was rooted in efforts to map and chart the wilderness of human taste so that soldiers' bodies could be kept healthy enough to wage war.[23] However, measurement of preference, or sensory thresholds—the point at which one can barely perceive something—failed to capture what

Nadia Berenstein refers to as the "content of flavor experience . . . the interrelated chemical and perceptual changes that could occur with a single alteration to the composition of a food product."[24] Measuring sensory experience presented a scientific and technical challenge that cut across disciplines and techniques.

Among the many techniques discussed at the symposium was one called flavor profile analysis (FPA). Workers at the Arthur D. Little Company (ADL), a consulting company based in Boston, had developed FPA in response to difficulties encountered while examining the sensory impact of adding monosodium glutamate (MSG) to foods, especially foods with multiple types of seasonings.[25] Findings suggested that MSG worked best below the perception threshold in accentuating flavor, and that it could suppress unwanted flavors such as the earthy and peel-like flavors found in mashed potatoes. Like other behavioral analysis techniques being developed at the time, FPA took individuals, with their individual ways of perceiving and verbalizing flavors, and trained them to develop a shared way of talking about the tastes they encountered.[26] FPA sought to translate the "art of flavor impressions" into a "scientific conception."[27]

First introduced at the 1949 meeting of the Institute of Food Technologists, FPA claimed users could scientifically capture and describe the "natural processes (often performed unconsciously) of evaluating and comparing flavors by describing their impressions."[28] To do this, panelists had to develop a vocabulary rooted in shared experience. Panelists learned to identify the taste, odor, or flavor of a substance using the same words in relationship to a reference. They practiced assigning the same intensity value to standardized dilutions. This required that panelists, despite their differing physiological abilities to perceive the intensity of flavorful molecules, train their bodies to consistently rate a standardized sample determined to have an intensity of, for example, "9" as a "9."[29] Through the usage of smell or taste "standards" that all participants could experience at the same time, FPA opened the doors to the shared experiences necessary for developing a vocabulary. The aromas and tastes identified during the process can then be linked to already existing knowledge about the flavor characteristics of known molecules.

FPA not only sought the ability to capture impressions, it also sought to replace the individual expert with a group. Doing so shifts the type of knowledge produced. There are two categories of expert knowledge that are useful to think with here: (1) connoisseurship—individual, embodied knowledge gained through repeated exposure—and (2) a panel of experts, where expertise is distributed across multiple bodies. ADL's employees argued to potential clients that relying on any single individual's expert knowledge was more risky and less reliable than the knowledge gained from a group.[30] FPA mitigated risks, proponents noted, by spreading the techniques of knowledge production across multiple bodies. This meant that if one person retired, left the company, or unexpectedly died, others could step in and easily replace the original expert. The move toward distributed expertise in sensory analysis, rather than individual connoisseurship, reflected trends occurring in many arenas of industrial production throughout the twentieth century. FPA imagined people's bodies as fungible, able to be swapped out one for another in a manner reminiscent of the logics of assembly-line production.[31]

This is not to say that those who were part of a sensory panel were just like anyone on the street or in the office. Even though one did not need to be an expert to join a panel, to participate in FPA potential panelists had to first pass a series of tests evaluating their ability to taste and smell "normally" as well as demonstration of normal intellect and intellectual integrity.[32] Prospective panelists who passed those tests were then trained in the techniques through seminars and practice sessions for up to a year. This training not only focused on minds, it also enrolled bodies, in the work of standardization characteristic of twentieth-century sensory labor by teaching participants to smell and taste in new ways: short "bunny" sniffs came to replace long inhalations, a pause between smelling and tasting became normal, liquids were slurped instead of sipped, and bite size was standardized.[33] Becoming part of a FPA panel meant becoming someone whose body and brain could step aside from likes and dislikes to focus instead on measurement.

Imagine a meal consumed this way! We see here an inversion of how people normally eat, a slow tasting with conscious attention not only to flavor, but also to texture and mouthfeel rather than the often less conscious and more social act of eating. This inversion, like the other forms of sensory testing explored in the previous chapter, relied on remaking how people used their bodies to perform everyday tasks. The differences were stark: one spat out the thing tasted instead of swallowing it, one munched on a dry cracker or sniffed wet cardboard in between tastings instead of moving immediately on to the next sample, and one sat alone in a booth or in a group at a round table crowned with a rotating tray.[34] All of these tweaks to the tasting process highlighted the difference between everyday and expert panel modes of tasting for both the tester and the taster.

In contrast to difference tests that measured whether a taster could distinguish two samples from each other, FPA relied on the development of consensus around the measured flavor's components. FPA's proponents posited that by using a consensus approach, the technique could erase individual likes and dislikes. They argued that FPA instead produced a collaborative numerical and linguistic description of perception. During initial sessions, panelists were given a food to individually study and taste. Through smelling, tasting, describing, and comparing the food in question to similar foods, each panelist would generate descriptive terms for the flavors encountered.[35] As panelists initially tasted a sample, their way of sensing the world temporarily transformed into an expert way of sensing, albeit only for one specific product. This expertise ideally vanished, the protocol designers argued, when panelists assembled together under direction of a group leader to reach a consensus on the sensory profile of the tested food. FPA situated expertise in a group rather than an individual.

FPA took time. It required that the group work together to align individual perceptions. Over a series of sessions, the panel not only needed to identify a mode for presenting *and* investigating the sample in question, they also had to collaboratively develop a range of tools. These included identifying a vocabulary linked to reference standards; putting in place protocols for tracking the information produced; deciding which flavor factors identified were important to the flavor and to what extent; and finally, identifying, defining, and agreeing on a frame of reference for the flavor "amplitude" of a product.[36] Although the steps appeared straightforward when presented in promotional brochures or on paper, the reality of producing shared experience required overcoming a much trickier hurdle: transforming smell and taste into words that, like the TON developed by Baylis and his colleagues, could be put on paper in a form capable of traveling beyond the laboratory.[37]

The goal of describing a person's sensory experiences through words relied on a specific, individualized understanding of material identity. "Since each material possesses flavor and aroma characteristics which in type and intensity distinguish it from any other material, great emphasis is therefore given to exact descriptive terminology," proclaimed the introductory publication about FPA.[38] During the 1950s, FPA's developers sought to walk the tricky line inherent in characterizing flavorful foodstuffs; they knew the molecular identities of some compounds that caused off-flavors, but not all. Complex mixtures like foods further complicated things! Nonetheless, FPA's developers, and the larger community studying flavor, understood materials as possessing unique flavor and aroma characteristics that revealed themselves during the process of tasting and smelling. As such, the development of flavor profiles sought to account for how known sensory inflection points behaved in complex mixtures. Flavor profiles helped quantify the impact of otherwise uncharacterized sensations. FPA offered an easy way for product developers and sensory scientists to visually communicate to upper management how product design decisions would impact flavor. The method's champions claimed it could help product developers' work regarding what they all too often didn't know: the makeup of the individual molecules in a complex mixture.[39]

## MEASURING MOLECULES

Understanding the characteristics of individual molecules in mixtures was seen as the holy grail for manipulating flavor. In 1951 consultants from ADL submitted a thirty-two-page report to the Bristol-Myers company detailing the effect of different ingredients and formulations on the overall flavor of Ipana toothpaste.[40] Recent batches of the toothpaste, ADL reported, had "weedy, garbagey, rancid, minty, spicy, soapy, bitter, and cooling" flavors.[41] In contrast, tubes of Colgate purchased that same year had a simpler—and notably less offensive—flavor profile. Drawing on results obtained using FPA, the report warned that *all* the toothpaste's ingredients could introduce off-flavors or undesirable odors: glycerin, stearic acid, soaping agents, thickeners, the essential oils used for flavoring, and even the toothpaste tubes threatened the product's overall quality and acceptability. Although not explicitly stated, comparisons between Ipana's flavor profile and that of its chief competitor, Colgate, hint at Bristol-Myers' concern with burgeoning competition for access to the mouths and dollars of consumers. The report argued that using "odor- and taste-free base materials" as well as obtaining flavoring oils "with as few off-notes as possible" would help Ipana obtain "flavor leadership."[42] To get there ADL suggested aiming for "blended flavors"; consumers would perceive such flavors as a "well-landscaped" experience.[43] "Keep close control on the quality of all the flavoring materials, as well as the quality of the material for the base, especially to insure that they be as tasteless and odorless as possible," consultants recommended.[44] Business success, their communication tools suggested, lay in mastering the molecules shaping consumers' olfactory and gustatory experience.

ADL promised increased understanding of how flavorful compounds behaved in foods. Yet understanding the characteristics of molecules had up until the 1950s been primarily the work of perfumers and flavorists. This began to change in the mid-twentieth century with the "analytical revolution" in the types of instruments available to chemists. Instruments allowed chemists to go beyond characterizing a molecule; instruments helped chemists identify the core components and behaviors of a molecule.[45] Being able to pinpoint the makeup and behavior of molecules fundamentally changed how chemists understood substances: molecular structures that had once been *properties* of a substance identified through tests became *objects* available themselves for classification.[46] In other words, molecules became things that could be synthesized, identified, classified, named, and regulated, just like any other object.[47] Taken together, instrumentation and sensory analysis pulled the art of flavor manipulation into the world of science; the two techniques opened the door for a new mode of understanding the world of scientifically uncharacterized, but nonetheless perceptible, materials.

For perfumers and flavorists, the "uncertain composition" of smelly mixtures remained a problem, whether they were making perfumes or manipulating the odors found in toothpaste, cleaning agents, or candy bars.[48] The new tools emerging as part of the analytical revolution in instrumental chemistry offered the possibility of understanding what was in a mixture. Of all of the instruments involved in changing how chemists understood molecules, gas chromatography (GC) proved most useful. GC builds on the principles of chromatographic separation: the idea that the physical properties of a substance—its weight, the elements it was made of, whether it dissolved in water—determined how quickly the substance traveled through a medium. Traditionally, chromatographic methods relied on using a liquid (aqueous) phase to separate out mixtures. Unfortunately, it can be difficult to suspend smelly molecules in a liquid; they evaporate easily! According to physiologists, molecules that easily evaporate, that are volatile, are the molecules smell receptors in the nose and throat detect and evaluate as having an odor. The very characteristic that makes it easy for someone to smell something in turn makes analysis of smelly molecules difficult.

Unlike traditional chromatography, GC replaced the liquid carrier agent with a gas. This meant that compounds existing in the vapor phase, like odors, could now be separated out into their individual molecular components rather than remaining complex mixtures. GC took the ephemeral and made it something that could be defined. Further refinements, such as development of various detection devices, facilitated characterization of the purified, eluted components. When researchers in 1959 coupled GC with mass spectrometry—an instrument that chemists use to determine the chemical structure of a molecule—they gained additional insights. The two instruments together allowed chemists to separate out volatile compounds in a mixture and to identify structural information about the individual molecules in the mixture.[49] These new technologies made it possible to "peer" into the previously unseen molecular world, to perceive a previously imperceptible granularity in the symphony of olfactory sensations caused by mixtures.[50]

Ten years after the introduction of GC, one chemist excitedly noted in a textbook that GC "offers every hope of allowing future study of problems hitherto considered too complex even to contemplate seriously."[51] Although the technology, as well as sample preparation techniques, remained in constant flux, by the early 1960s researchers from academia to industry were publishing papers based on GC work. The instrument, they reported, was useful for determining the molecular makeup of the natural world. Even better, it also allowed tracking of biological reactions and facilitated quality control in manufacturing. In short, researchers understood GC as the hot new technique for unlocking nature's secrets.

Yet for those working on issues of off-flavors or odors during the instrument's first decade, GC was not the perfect instrument for *all* disciplines. The questions it answered for an oil chemist were not the questions posed by a flavor chemist, and those were not the core concerns of a flavorist. As one scientist noted in a lecture delivered to the Society of Cosmetic Chemists in 1961, "The trained nose is a detector of a far higher degree of organization than any known gas-liquid chromatography detector. If geraniol, for instance, is fed into a gas-liquid chromatography detector, the response tells only that *some* material is there. By contrast, the perfumer's nose yields the identification 'geraniol,' indicating the structure down to stereochemical details!"[52] The division between the efficient "objectivity" of the machine and the subjective human nose mattered. GC could not offer those trying to manipulate flavors and odors the precision of a trained nose. It could not organize and categorize odors the way that an expert "smeller" could. In contrast, noses failed to "resolve mixtures into separate lots, each composed of molecules of a single kind" for independent examination.[53] Noses, for all their specificity, struggled with the work required of them by an industry interested in minimizing consumer reports of flavor problems.

## SENSING AT DISCIPLINARY CROSSROADS

Unlike the problem of an excess of tastes or smells faced by the MWD, many industrial food producers struggled with flavor loss. For example, researchers working on managing the quality of canned pears found that canning resulted in "volatiles, including many that contribute to flavor, [being] driven off to the atmosphere."[54] This meant that the cannery smelled amazing, "blanketed in a rich, desirable aroma."[55] Unfortunately, the resulting product "almost totally lacked flavor."[56] Rose Marie Pangborn, a pioneering sensory scientist at U.C. Davis, led the researchers investigating flavor loss. Pangborn was interested in understanding flavor across multiple mediums, from pears to wine and water. Pangborn and her colleagues linked the dreaded flavor problems of industrial production—blandness, taste of packaging, and rancidity, to name just a few—with loss of volatiles caused by the high levels of heat and pressure often used to extend shelf life.[57] Despite claims to exceptional flavor quality from perfume and flavor houses, variations in stock flavors further complicated producers' efforts to manage unwanted tastes and odors. Facing fierce competition for grocery shelf space, food companies and researchers understood that consumer dissatisfaction threatened market share as much as the success of transforming fragile crops into foods that could last. For industrial producers, success in their self-appointed mission to feed the hungry world of the future, as well as in the marketplace, was intimately entwined with sensory management during and after production.

Notwithstanding the excitement about GC, in the decade after its introduction the now "routine" technology's limits became apparent.[58] Although GC measurement helped researchers better understand what compounds were present in food, those same measurements failed to explain what impact a specific flavor molecule might have. To address how bland, boring canned pears could emerge from an olfactory paradise, the researchers at Davis and their industrial collaborators decided to investigate which of the lost pear aroma components actually mattered to sniffing noses.[59] Rather than an instrumental detector, they asked a group of trained judges, selected for their olfactory skills, to smell sample fractions obtained from the GC.[60] The judges graded the resulting fractions for pleasantness/unpleasantness and aroma weakness/strength; subsequent testing included providing "detailed description of all odor characteristics."[61]

Like FPA, the approach employed by Pangborn and her colleagues relied on group, rather than individual, expertise. The researchers understood that humans detect sensory signals differently than instruments: Noses don't just detect; they react. As the judges sniffed samples, their bodies sorted each sample. The results allowed Pangborn and her colleagues to identify which fractions mattered, which in turn helped the researchers isolate and purify aroma extracts to add back into the cooked pears. In indirectly joining the nose with the separating power of the instrument, the researchers attempted to rewrite the problems of industrial processes.

Despite efforts like Pangborn's, on the whole researchers were accused of lopsidedly using GC. "There is no doubt that some important components contributing to the aroma of foods have been isolated and identified. *What is usually not being done* [by GC researchers] is to establish *which* of the compounds isolated are responsible for *what* sensory properties!" complained George F. Stewart in 1963. Editor of the Institute of Food Technologist's two major publications, *Food Technology* and the *Journal of Food Science*, Stewart worked across industry, government consulting, and academia.[62] Given industrial efforts to improve, enhance, or simply restore flavor to food, it is no wonder Stewart found GC's promise a disappointment: "It seems likely that many of the compounds isolated have little or no sensory effect at all" he noted.[63] Stewart encouraged researchers to integrate the body's reactions with GC investigation.[64]

Stewart was not alone in his views. Emily Wick, the first woman to receive tenure at the Massachusetts Institute of Technology, noted in a 1966 piece for *Food Technology* that "the outlook for flavor research is very good insofar as the 'chemistry' is concerned." But, she warned, when it came to flavor, "significant work remained."[65] For both Stewart and Wick, the scientific way of knowing the world—identifying and cataloging the components found in foods—failed to align with the discipline's self-appointed task of bringing "man's needs and industry goals" together, as the cover of the November 1966 issue of *Food Technology* suggested. Stewart and Wick reiterated that food scientists carried the responsibility of thinking of how molecules mingled with bodies and shaped behavior; doing so required pairing human and machine.

## AROMA ROAD MAPS

Perhaps it is not surprising that the innovation that addressed Stewart's and Wick's critiques came from industry. After all, industry is the space where market forces most closely intersect with human sensing, from car design to package color. The setting was fairly anodyne: three Colgate-Palmolive employees in the mid-1960s were trying to understand how two samples that appeared identical via gas chomatographic analysis could smell so different.[66] Curiosity piqued, they explored what would happen if they replaced the electronic sensing device located after the GC column with their perfumer colleague's nose. Smelling the helium stream as it exited the GC was not new, the authors noted. What, they wondered, could be learned by using an expert human detector who carries an olfactory encyclopedia in their brain?

Their results were exciting. The perfumer connected smells with smaller peaks that only appeared on the spectrum at maximum instrumental sensitivity. These peaks, generally undetected due to lower sensitivity settings or ignored due to their size, turned out to contribute to the difference in smell! With the machine's help the perfumer's nose successfully separated out "fellow travelers" from "pure aromatics."[67] Even better, due to repeated training with pure extracts, the perfumer could identify the chemical composition of some of the smells that came off the column. The perfumer validated the instrumental detector and vice versa.

By pairing perfumer with chemist and machine, the researchers at Colgate-Palmolive demonstrated that it was possible to learn "the relationship between pure major and minor components." Furthermore, one could identify the influence of minor components on overall olfactory experience.[68] Bodies and instruments carried different capabilities. Researchers at ADL, for example, reported that in some cases a human nose identified differences in odor quality between pure and impure compounds; in contrast, GC paired with an instrumental detector signaled the compounds were the same.[69] Human, 1; machine, 0. At the same time, the researchers noted, the (nonexpert) human nose did best at measuring compounds. Individual components proved trickier. "We feel that only by a cooperative effort combining the two techniques [smelling and GC] can we obtain definitive information regarding odor perception and the stimuli," the ADL researchers concluded.[70] Used together, bodies and analytical instruments opened the door to linking odor descriptions with molecular identity. In the process, GC opened the door for changing understanding of perception to include the idea that every smell could eventually be linked to a specific molecule(s).[71]

As Baylis, Gullans, and Spaulding noted in the 1930s, developing a shared language for talking about smells and tastes was not easy. Despite the promise offered by using human noses as detectors for gas chromatographs, a lack of standardized communication tools made it hard to reproduce or verify research results. The Colgate-Palmolive researchers worked around this by measuring how long it took for a molecule to travel through the chromatographic column and be detected by the human sensor at the other end. As proof of concept, measuring the amount of time for a molecule to travel through the column worked. It helped show that an expert could detect trace amounts of odorific compounds, especially when separated out rather than in a mixture. It even verified that their perfumer colleague could do what he claimed: use his nose to identify molecules in really small amounts. However, retention times (the time it took for a molecule to travel through the column) were only relative to the known standards used. This meant that any other researcher interested in using the retention times reported by George H. Fuller and his collaborators would have to *exactly* reproduce the conditions of analysis. Without a standardized mode of communicating retention times, chromatographers found themselves in a situation not all that dissimilar from that of waterworkers in the 1930s: their locally produced knowledge stubbornly remained tied to the places and conditions of its production.

Communicating about retention times wasn't the only challenge to adopting human noses as sensors. Perfumers cost a lot; they were possibly "the most expensive detector ever proposed for a GC."[72] They also carried a much larger olfactory vocabulary than the everyday person. To be useful in flavor problems, any tool linking molecular identity with retention time and perceived aroma descriptor needed to be usable by specialized researchers *and* technicians with little training—it needed group expertise.

Just as concern with flavor problems crossed disciplinary boundaries, so too did solutions for the challenge of communicating about research findings. Manufacturers of gas chromatographs, aware that their customers wanted better communication tools, worked to help promote adoption of a generalizable mode of calculating retention times. The same year that Fuller and colleagues published pictures of their perfumer standing in a phone booth and sniffing the GC effluent, Leslie S. Ettre, the scientist tasked with promoting Perkin-Elmer's new GC published a paper in *Analytical Chemistry* encouraging adoption of the "widely accepted in Europe, but virtually unknown in the U.S." Kováts Retention Index.[73] Ettre used a whirlwind tour of workshops, presentations, and publications to promote the retention index to academic, governmental, and industrial users; researchers in all three arenas widely adopted it.[74] Retention indexes transformed communication; they made it easier to dissociate sensory experience from the body and easier to develop a shared understanding of what "qualities or 'notes'" applied to common molecules.[75] Adoption of retention indexes, facilitated in part by GC manufacturers, allowed locally produced measurements and individual analytic skills to circulate far beyond the laboratories where they were produced.

Retention indexes weren't the only solution for sidestepping time-intensive, routine sensory evaluation or the expertise embodied in a perfumer. Innovations in sample preparation also promised to help. Understanding olfactory experience, researchers argued, meant more than analyzing chemical *extracts* from a food; as the U.C. Davis researchers had demonstrated with pear aroma, understanding smells required analysis of the vapors coming off the "headspace" above a food. Unfortunately, many of the compounds that are aromatically significant exist in really small amounts. Headspace analysis of such compounds would require massive sample sizes! So how to do this? Andrew Dravnieks and Anne O'Donnell, researchers at the Illinois Institute of Technology Research Institute's Odor Research Center, suggested a dual approach. By first using chemical extraction and mass spectroscopy to identify molecules, they could then use headspace analysis via GC as a "road map for locating, concentrating, and isolating components significant for aroma."[76]

Maps make it easy to navigate unfamiliar territory. In using the term "road map," Dravnieks and O'Donnell highlighted how much researchers and food producers wanted to "bridge the gap between identification and flavor significance."[77] Dravnieks and O'Donnell expanded Fuller and his colleagues' approach outside of the expert realm by combining the use of retention indexes with the now "rather common" practice of sniffing.[78] Retention indexes and sniffing the air coming off the gas chromatographic column offered a rational approach to "locating and cataloging the components which carry particularly characteristic or strong odors."[79] This allowed scientists to concentrate their work "on those components which most significantly contribute to the full odor complex."[80] The immaterial and ineffable experience of smelling something delicious, it turned out, could be measured.

Thinking of the nose as attuned, interested in effect and affect, elevated the importance of the embodied response of the human detector in researchers' eyes. However, this process also highlighted the challenge for nonperfumer researchers. How could they communicate their own olfactory encounters with smelly molecules exiting the GC column, given how difficult it was to reliably tell each other what they had sensed?

Learning to articulate sensory effects called for a standardized vocabulary that could, like retention indexes, travel across disciplinary boundaries. In the 1970s, fifteen labs working under the Sensory Evaluation task group of the American Society for Testing Materials (ASTM) began collaborating together to "stabilize" semantic methods for characterizing odors building on previous work.[81] Semantic methods assign a descriptive word to an odor, then rate the words for whether they fit.[82] However, as Dravnieks noted, when multiple people used semantic descriptors, the results became a "noisy" mess despite the ease of using words to describe an odor as opposed to the alternative of rating the odor's similarity to a reference.[83] To determine whether the noisy mess could be stabilized, the research team collected more than 830 already in-use odor descriptors. They then screened the list, trimming it down to 160 words. Finally, the team tasked 150 people with evaluating the suitability of each of the 160 descriptors for ten samples on a scale of 0 to 5.[84] Participants evaluated each odor twice, with evaluations separated by a few months. By aggregating participants' results by "either the frequency or the sum of scores," the team developed a "percent applicability" method; this allowed them to demonstrate that profiles developed by a large number of panelists, when aggregated, produce "stable representations of the odor character."[85]

Many of the words in the descriptor list easily connected to items found in the environment, like *soapy*, *leather*, *cardboard*, *rope*, and *wet dog*. Others, such as *bitter*, *sharp/pungent*, and *putrid/foul*, overlapped with culturally informed definitions of smells. ASTM hoped that adoption of standardized descriptors would make communication about the smell of molecules more reliable, capable of circulating *despite* the generally subjective experience of naming odors. Over the next two decades, this approach of linking flavor/odor descriptors with retention times, and thus molecular identity, became an increasingly common part of conducting analysis.[86] It also sidestepped the "problem" of expensive perfumers.

The codified combination of a human nose—expert or otherwise—with machine was an inflection point in how researchers understood smell and tastes. It provided the granularity desired by waterworkers trying to navigate unwanted tastes and odors. In promoting standardized modes of communication, manufacturers and private researchers paved the way for describing olfactory experiences through biochemical terms. To riff on Sophia Roosth, they "molecularized" smell.[87] Standardized modes of communication promoted adoption of instruments coupled with humans to measure volatile molecules. The subsequent success of using human and machine created a new epistemic culture: researchers understood that smells and tastes signaled the presence of molecules.

Like a similar move toward the molecular in the life sciences, the combination of human and machine-facilitated efforts sought to optimize the future by acting in the present on molecules as they were identified.[88] This shifted the boundaries of expertise around sensation: no longer was the expert only understood as a connoisseur who held knowledge inaccessible to most. Instead, the expert's identity expanded out to encompass analyst-technician, aroma glossary, and scientific machine. In the process, odors previously considered difficult to characterize, such as those found in naturally occurring products like essential oil extracts (central to most perfume and flavor work), became accessible, understood as little bits of malleable data.[89] Gaining the ability to control tastes and odors seemed just a matter of money. And time.

## TASTING WATER LIKE A FOOD

Industrial pollutants like phenol defined many of the taste and odor problems that Baylis, Gullans, Spaulding, and colleagues described in the interwar period. By the time MWD's McGuire faced a diver with hands full of smelly green goo, waterworkers had spent almost three decades working to understand taste and odor issues caused by post–World

War II agricultural practices.[90] Chemical fertilizers in agricultural runoff had increased growth of algae, phytoplankton, and other microorganisms in many surface waters. Scientists first associated microorganisms with the production of earthy odors that could "taint" waters (and the fish that swim in them) in the 1930s.[91] It took another thirty years before N. Gerber and H. Lechevalier identified the specific compound responsible for earthy odors; In 1965 they published a report identifying a compound they named geosmin (ge = earth; osme = odor) as the source of earthy smells coming from a group of bacteria known as actinomycetes.[92] Four years later, two groups independently identified and isolated a second compound with a musty odor, 2-methylisoborneol, also produced by members of the actinomycetes family of bacteria.[93] Subsequent research has demonstrated that additional players—from industrial waste treatment facilities, drinking water treatment plants, and local ecosystems—contribute to the formation of geosmin and 2-methylisoborneol.[94] To date, researchers note that eutrophication, the lack of oxygen caused by human activity in some form, seems a significant producer of geosmin, and 2-methylisoborneol occurs via a combination of human-caused and naturally occurring microbial activity.[95] In short, scientific understanding of geosmin and 2-methylisoborneol is still incomplete.

Between the odor episodes of 1979 and 1980, workers at MWD adopted a new analytical approach.[96] The technique, closed-loop stripping, offered sensitivity "at or below the lowest threshold odor concentrations" reported for major odor contaminants. Workers hoped it would help improve their ability to target treatment.[97] It did. Unfortunately, its success was only partial—despite insights from the new technique, despite the additional treatment steps put in place to prevent additional problems, a small number of complaints continued to trickle in.

The complaints, it turned out, were caused by very small levels of geosmin (3-5 ng/L), and methylisoborneol levels at *or below* the instrument's detection level (2 ng/L).[98] A subset of the population was sensitive enough that their noses could detect the molecules causing the earthy, musty odors at levels below that of the most advanced analytical instrumental approaches of the time!

Faced with the continued, albeit low-level, presence of musty smells, McGuire suggested the MWD follow advice published by two of his graduate school mentors in 1971: that waterworkers borrow approaches used by the perfume and food industries.[99] Informed in part by McGuire's exposure to FPA while a graduate student at Drexel, the MWD invited consultants from ADL to train local waterworkers to use FPA.[100] The epistemic culture of flavor as molecular that had emerged over the previous twenty years suggested that successfully responding to flavor problems required more sensitive detection techniques and the ability to increase the granularity of available data about tastes and smells.

Although first aimed at the food industry, by the 1970s ADL began actively marketing FPA as a tool to "objectively measure and control odors" found in the environment.[101] "With the sharp increase in the public's concern with pollution in recent years, and the tendency to protest louder than ever, many companies which have operated without apparent problems for years now find themselves harassed by odor pollution complaints," one advertising brochure noted.[102] A modified version of FPA, coupled with "sophisticated analytical techniques," made the perfect tool for dealing with odor complaints.[103]

Abandoning the TON in favor of FPA changed who participated in the process of evaluating water. Rather than focusing on scientific expertise in any specific area, FPA called for selecting people with "moderate to high sensitivities for miscellaneous odors."[104] This brought together a variety of staff members. Microbiologists, chemists, and technicians came together for training on the methodology; those trained continued to serve as panelists for the remainder of their careers. Unlike TON, FPA used consensus and focused on development among panel members of a shared language *about* tasting. Repeated testing sessions maintained the "muscle memory" of tasting not only by reinforcing the physiological link between odor and memory, but also by reinforcing the shared language developed by the panel.[105] By 1982 the MWD was using FPA to catch and remedy flavor problems before the water reached consumers and to solve taste and odor problems when conventional chemical analysis failed.[106]

Molecular compounds such as the earthy-musty smelling geosmin and 2-methylisoborneol from actinomycetes and other bacteria have existed in water for years. Chlorine masks the subtle sensory inflection points where geosmin and 2-methylisoborneol activate perception. Yet as water travels away from the production plant, chlorine dissipates. People living at the end of distribution lines carrying water from the MWD in the early 1980s were much more likely than those living close to encounter earthy-musty odors in their glasses of water.[107] Sensory inflection points, perceived, resulted in consumer complaints. By integrating FPA into its detection and treatment approach, the MWD increased the granularity of data at the plant in a way that made it easier to detect and act on geosmin and 2-methylisoborneol in water long before it ever reached a consumer's glass.

The ability to identify the compounds causing musty and earthy odors rested on an increasing granularity of data formed through a combination of instrumental innovation and human-produced sensory insights. Inflections in water's tastes and smells joined with technological innovations to shift the pathway followed by US waterworkers in identifying and acting on off-tastes and smells toward one that used humans and advanced instrumentation in tandem to identify and remove tastes and smells before they reached consumers. At the same time, perceived sensory inflections—such as when a compound like 2-methylisoborneal could be detected by humans, even if instruments missed it—occasionally revealed otherwise imperceptible human-environmental interactions. For example, agricultural runoff from intensive farming practices can exponentially boost algae growth, which in turn can increase the presence of earthy or musty odors. As such, earthy or musty odors in water not only tie drinkers back to the weather or season, they also (but not always) point toward the impact of other human activities on water quality.

In subsequent years, water treatment professionals in the United States came to widely accept FPA as a useful method for addressing off-tastes and odors. The epistemic culture that made flavor molecular in the mid-twentieth century resulted in a convergence between how food and sensory scientists and the engineers and chemists working in municipal water production analyzed and evaluated unwanted tastes and smells. Internationally, many water treatment professionals faced similar challenges. Dealing with the unwanted tastes and smells caused by degraded water quality, be it from industrial pollution or agricultural runoff, required a reorganization of how individuals and institutions approached the molecules that triggered sensation. As the next chapter shows, addressing unwanted tastes and smells pushed waterworkers to form new relationships that crossed linguistic and national borders. Their work would redefine long-standing hierarchies of tasting.

# 3

# FUTURE SENSING BODIES

IN JULY 1970, *L'Express*, a French lifestyle magazine, reported a new system of monitoring water quality in Paris. "Water quality," the report noted, "is the affair of a man, largely unknown to the public, the water taster." With a "virginal palate that eschews nicotine and spices," the water taster "finds joy in noticing nothing at all." Tasting water warmed to 30°C happened hourly, the temperature chosen as the optimum detection threshold. "He tastes, gargles, gurgles, and spits. Then he smiles contentedly because he has sensed the same insipid sensation, colorless, odorless, that he noted the previous hour, and the day before, and the day before that, and the years before."[1] The water taster, the author of the article in *L'Express* intimated, had carefully protected his mouth and nose from damage, sacrificing the normal hedonistic pleasures of French life to preserve the boundaries society and science deemed necessary between raw and "cooked" water. As a gustatory sentinel, the water taster used his mouth and nose to detect subtle inflections in treated water before it entered the distribution system, intentionally monitoring water's perceptible markers and inviting action long before it went out to encounter the bodies of consumers at the taps throughout Paris's distribution system.

*L'Express* offered readers a vision of the sort of technical and gustatory expertise that Baylis, Spaulding, and their colleagues had anticipated more than three decades earlier, albeit with the distinctly French twist of tasting water like wine. Paris's water tasters had been asking "Does it taste good? Does it taste like water?" (as one Baltimore reporter cheekily put it) for at least a decade.[2] Unfortunately, as the article in *L'Express* pointed out, local consumers were "not persuaded of the water's quality": many "well-informed" residents in the tony 16th arrondissement obtained their table water from a fountain in Lamartine Square fed by an artesian well even though chemical and gustatory analysis demonstrated no superiority to tap water; similarly, "54% of Parisians will only drink mineral water," despite the price being a thousand times higher than that of tap water.[3]

*L'Express*'s brief discussion about how water tasters helped ensure Parisians received water that tastes "like water" came at a critical crossroad in French water management. Publication of selections of Rachel Carson's *Silent Spring* in the popular journal *Paris-Match* in 1963, as well as a subsequent complete translation by famed publishing house Plon, had highlighted to the public that human-made chemicals in the environment harmed animals and people alike.[4] Despite increasingly widespread public awareness of the potential molecular dangers to the health of individuals and ecosystems caused by industrial pollution, as of 1970 French law did not restrain industrial factories from "changing pure water into a torrent of trash."[5] The author of the article in *L'Express* noted that in the Parisian region only a third of wastewater received additional treatment before discharge: "The rest, like everywhere in France, continues to pollute the rivers."[6] Unlike Paris, many medium and small French towns in 1970 were struggling to provide adequate water for inhabitants and industry. Pollution, a weak regulatory regime, and the injustice of a capital city with an excess of water juxtaposed against a thirsty countryside rendered French water suspect. The production of good tasting water, the article suggested, needed to be accompanied by a strong national regulatory regime capable of, at a minimum, overseeing water use. The article argued that a national organization tasked with inventorying and recycling France's waters while fighting pollution would be even better.

Years of accumulated industrial discards, recent adoption of synthetic fertilizers in agriculture, and a new political turn to addressing aging infrastructures through privatization came to define efforts around managing the tastes and smells in municipal water in the 1970s–1990s. As private water providers expanded their business aspirations to a multinational scale, the ability to quickly respond to taste and odor problems in locations far from the centers of technical expertise gained importance, putting in place an inflection point in how engineers and chemists working in municipal water in the United States and France thought about consumers. A business-driven need emerged for understanding how to manage the relationship between tasting bodies in the here and now and the future sensing body of an aggregate consumer. In the process, regional geographic nuances in the local terroir of water increasingly verged toward a more standardized aesthetic quality.

## TASTING A CRISIS OF QUALITY

The article in *L'Express* reflected national concerns about water pollution that had swirled through France's social and political spheres throughout the 1960s. Such concerns led to legislative action that increased protection for French watersheds in 1964, although it would take another three decades for additional clarification to emerge, as well as the emergence of France's Green Party.[7] A drought in 1976 further foregrounded questions of quality and access to water. Josef Olivesi, the general engineer of bridges and roads, remarked to reporters on June 18, 1976, that dryness had resulted in further degradation to the quality of water producers were treating. "For each cubic meter of water to treat, we must eliminate a much greater quantity of pollution," Olivesi noted. He continued, "We can guarantee in all circumstances that we will deliver to the population a potable water in the sense regulated by the Ministry of Health; but we cannot absolutely guarantee that this water will be perfect to the taste."[8] Producers of Paris's municipal water knew that their ability to manage the flavor of water coming out of taps depended on a combination of the quality of the water coming into their production facilities and their own technical skill set. Quality troubles meant trouble with taste.

As citizens and politicians became increasingly aware of the link between pollutants and water quality, public opinion and political will demanded solutions. Olivesi was painfully conscious of the gustatory failures of Paris's municipal water when he spoke to reporters on that hot June day in 1976. He promised potable water, but he did not guarantee that Parisians would enjoy drinking it. Taste and smell, no matter the measured safety of the water, mattered; *Le Figaro* noted in December 1978 that although Paris's water, chemically purified through filtration, chlorination, and ozonation, was safe to drink, "the consumer requires that their water also have an agreeable taste."[9] It was to this end that the city's municipal water producer had employed water tasters. Dressed in white shirts, a visible sign of the purity their sensitive taste buds would (ideally) detect as they tested water, these workers offered salvation from Paris's aquatic woes. "As soon as they detect the smallest strange taste [the tasters] intervene. Chemists quickly conduct new analyses, delivery is temporarily stopped, long enough to suppress the observed off taste."[10] In bringing water tasters into the technical toolbox used to make water, Paris's water producers situated taste and smell as the markers of quality that mattered most to consumers, markers that called for action in the here and now capable of minimizing any negative physiological reaction from a future consumer's sensing body.

## FUTURE SENSING BODIES

Those working on food—be it in the traditional sense of things one chews or this book's expanded definition that includes water—have a long history of trying to identify the causes of off-flavors and odors before foods get to consumers. They do this preventative work so that some body, located somewhere along the distribution chain, won't detect anything wrong at some unspecified point in the future. For example, early on in their training undergraduate and graduate food science students in the United States take part in a laboratory course in which they purposely produce and learn to detect off-flavors in food. One semester, early during my graduate food science training, I took a food processing class and lab. During one lab period we exposed milk samples to light, freezing, or extended room temperature. The treatments resulted in development of different, common, off-flavors in the product; remember the taste of milk in a carton from elementary school lunch? When my classmates and I returned to the lab for the next class, we brought worksheets, writing utensils, and our noses. Little sniffs of different, de-identified containers holding the "compromised" product helped us identify which products had developed off-flavors. Discussion of our results and further smelling of identified samples taught us to viscerally connect the painty smells found in some milk with too much exposure to light. Unpleasant smells connected past molecular interactions, invisible to our eyes, with different types of failures in the food processing and distribution chain. Through this exercise, my classmates and I learned that our noses and mouths could help us detect molecular interactions, and we could in turn take action to prevent those molecular interactions from occurring or identify when they had occurred.

Labs are a core part of training food scientists to understand the link between processing, storage, and consumer experience. As Ella Butler notes in her ethnographic examination of food science programs, training experiences like the one I participated in develop "a particular understanding of the consumer sensorium."[11] Consumers, similar to those responsible for developing and producing industrial food, are "highly sensitive to compounds at trace levels."[12] When food scientists or waterworkers think of consumers as highly sensitive detectors of trace compounds, they are doing so from a perspective that understands flavor as made up of molecules that will activate physiologic inflection points for one flavor experience, while other molecules might tamp down or eliminate physiologic inflection points for another flavor experience. This understanding of off-flavor, learned during the training process, recognizes that activation of inflection points at scale carries the potential to derail an enterprise's success. As I trained my nose in that laboratory exercise to recognize and name light-struck milk, I began to imagine consumers in the future as people like me: people who would pick up a glass of milk, sniff it, and even if they didn't necessarily name the problem, identify that the milk's quality fell outside of whatever they considered normal. I imagined a *future sensing body* and associated its future moments of tasting or smelling with actions taken all along the production and distribution chain.

The future sensing body I imagined as I tasted the light-struck or previously frozen milk is not any single body. Rather, the future sensing body is an aggregate of bodies sensing *at scale*. It comes into being when food producers extrapolate from present information about a foodstuff what changes might happen to food in the future. The future sensing body plays a significant role in how food scientists determine "best by" dates; it can upend business plans by disagreeing with instrumental measurements that claim a food is still of good quality.[13] While the majority of work dedicated to understanding and acting on future sensing bodies occurs within industrial food production, it occurs in artisan realms, too—as demonstrated by Harry West's examination of how cheese producers in the Auvergne region underage cheese so that traveling tourists eating it days after will experience deliciousness.[14] The future sensing body is always a projection into the future. It comes about through the official (and unofficial) compiling of the measured sensory thresholds, insights, and experiences of multiple bodies.[15]

Uncertainty helps bring the future sensing body into existence. This includes uncertainty about how any single individual body may sense—after all, every body experiences sensory unevenness due to causes such as illness, age, genetics, hormonal shifts, and environmental allergens. More critically, it includes uncertainty about the entire production system: each step (sourcing, production, distribution, consumption) shapes how producers imagine and seek to influence the future sensing body.

Industrial food producers can respond to uncertainty in ways that water suppliers cannot. For example, should a drought hit an ingredient supplier in one location, or a supplier in another location offer advantageous pricing, industrial food producers can back out of contracts or let contracts expire. Most ingredients are easily swapped out for a cheaper or more attractive alternative, as highlighted in the Ipana toothpaste case in the previous chapter (although as that case points out, not all peppermint extracts or other raw ingredients are created equal). In contrast, the raw water supplies municipal water producers rely on are relatively inflexible; built infrastructures and legal structures constrain water provisioning. Should a municipality wish to change its water supply from the single or multiple sources normally used, it cannot simply switch suppliers. Rather, the municipality must reroute pipes, navigate upstream water rights, adjust for different chemical makeups, and if they wish to be successful, adequately educate consumers on subsequent changes that might occur.[16] At the same time, the raw water supply remains widely variable. It changes daily, monthly, and yearly as rainfall, weather, and even upstream use shift. For water producers, the future sensing body activates concerns similar to those faced by the rest of the industrial food industry but requires action within a more constrained set of parameters shaped by what sustainability scholars refer to as the "locked-in" nature of built infrastructure.

Although uncertainty brings the future sensing body into existence, it is sensory science that makes the future sensing body actionable. Specifically, the work of determining the sensory thresholds at which detection and perception happen aggregates together insights from multiple sensing bodies. Information about the thresholds at which expert and everyday tasters and smellers detect (i.e., are able to identify) or perceive (i.e., are able to articulate) a difference makes it possible to act on the future sensing body. As explored in the previous chapter, instruments such as the gas chromatograph further assist in this process, enabling the identification of molecules capable of activating future sensing. When I sniffed light-struck milk and then responded to questions on a worksheet about steps that could be taken in milk production to prevent the appearance of painty or grassy flavors, I understood my sensory experience as a key step in preventing many other bodies in the future from having the unpleasant experience I had just undergone. Techniques such as adding antioxidants, flavor enhancers, or masking agents, and improving packaging or delivery methods each exist to ensure as much as possible that the future sensing body, *in its aggregate form*, will continue to accept the product for its predicted shelf life.[17] While an extremely sensitive body may reject a product, the future sensing body will accept it, preventing an unwanted inflection point at which mass sensing results in action that could undermine a company's fiscal success.

The idea of a water that tastes the same hour after hour and day after day is characteristic of industrial tastemaking. Indeed, the underlying premise of industrialization is that through scientific management of production one can work to overcome the vagaries of both nature and production processes. Success in this enterprise depends on projecting into the future what might be sensed. This desire drove Paris's water providers to hire water tasters in the 1960s and 1970s and inspired the MWD's adoption of the FPA approach in the 1980s. As the work of the water taster in France and the flavor profile panel in the United States demonstrates, the future sensing body does not only act to mitigate uncertainty. It links uncertainty with quality.

## THE PROBLEM OF QUALITY

The concept of quality straddles two worlds. In one, measurable characteristics as identified through instrumental tests and regulatory limits result in something being judged as a quality material. In the other, aesthetic judgments made by a connoisseur or expert (or even everyday expert) determine quality. When it comes to drinking water, producers and consumers may use measurable characteristics *and* aesthetic judgments to define quality; producers just sit more fully in the world of measurability, while many consumers make choices defined by definitions of quality that are shaped by their physiological and psychological reactions and thus fall more on the side of aesthetic judgment.

Western philosophers have argued that for something to be considered aesthetically, one must embrace a "disinterested" position, appreciating the object for its own sake, not because of some utility found in the object.[18] In this framing, people with "good taste" identify and judge quality outside of their own personal preferences. The word *taste* describes how capable one is of making judgments that align with social expectations; it also describes the physiological experience that occurs when tasting mouths meet taste-able things.[19] This "Problem of Taste," to use Carolyn Korsmeyer's framing, isn't new. Eighteenth-century European philosophers divided taste in two, separating the experiences our bodies have of tasting something (taste) from aesthetic judgments of Taste. Purposely distinguishing between the two tastes through capitalization, Korsmeyer highlights how for many philosophers, Taste came to be understood as "the ability to perceive beautiful qualities and to discriminate fine differences among the objects of perception, differences that might escape notice by someone without Taste."[20] Taste, quality, and aesthetics intertwine in ways that continue to defy measurement.

Initial efforts by psychophysicists to experimentally bridge the two worlds of quality focused on exceptional aesthetic experiences such as listening to music or looking at art. In the process, mundane and everyday experiences got shunted aside as unimportant to conversations about quality and Taste. Philosophers such as Korsmeyer and Yuriko Saito note that places such as the museum or concert hall frame the objects or performances within them as being worthy of aesthetic attention and value by those with good Taste. In contrast, Saito points out, everyday objects and experiences occur in places we are familiar with; as such, everyday things lack the framing locations that might mark them as worthy of aesthetic attention. "Our everyday aesthetic experience does not come to us in a neatly packaged bundle, consisting exclusively of qualities we receive through the 'higher senses' of vision and hearing as an uninvolved spectator," argues Saito.[21] Rather, Saito suggests that everyday aesthetic experiences draw all the senses together: smell, taste, and touch. Korsmeyer similarly points out that senses like taste, despite their traditional exclusion from the realm of the aesthetic, allow one to quickly and intuitively grasp an object's significance.[22] Everyday aesthetic experiences enroll people in a different relationship with the world: one of intimate, interested, *embodied* involvement.

Recall the water taster sitting in a room at a water plant in Paris, gargling, gurgling, and spitting out water. This does not sound like a particularly aesthetic moment. Recognizing the possibility of the aesthetic to show up in everyday experiences facilitates seeing the water taster's work in the Parisian lab as complementing the scientific work being done by engineers and chemists. The water taster managed the perceptible markers of quality, using his mouth and nose to mark embodied inflection points and in turn ensure the absence of an everyday aesthetic reaction in end users. In the process, his senses, and the senses of all attending to or ignoring the taste and smell of treated municipal water, sought to shape what could be known by future sensing bodies about the environment.

## MANAGING SENSING AS BUSINESS WORK

Funny tastes, smells, colors, textures, and temperatures help us evaluate how risky our environments are. Despite this centuries-old relationship, discoveries in the nineteenth century upended that relationship between sensation and perception of risk. As scientists linked invisible-to-the-eye microbial activity with disease, they also undermined the reliability of individual moments of sensation. Their research showed that water containing cholera-causing bacteria could taste, smell, and look perfectly fine, but drinking it could have devastating results. Aggressive public health campaigns, advertisements, media reports, and word-of-mouth stories about illness contracted from drinking suspect water taught a new lesson: drinkers needed to be leery of clear, good tasting waters for what they might hide. As germ theory gained ground, so did public health expectations that consumers of all classes would draw on scientific insights about imperceptible dangers as they used their senses to navigate the world around them. Rational drinkers were drinkers who knew to steer clear of untreated water, no matter how limpid. By the mid-twentieth century, *safe* water, defined by varying levels of chlorine wafting up to noses from glasses, became the norm in many places around the world. For those living in areas with robust municipal water systems, drinking water increasingly became an everyday activity capable of mass inattention.[23]

Unanticipated episodes of funny-tasting water complicate efforts to produce water that consumers like—or at least that they don't pay much attention to. Municipal water suppliers can predict seasonal variation and plan accordingly. For example, lakes deeper than four meters, such as Lake Mead (which, despite being located between Nevada and Arizona just east of Las Vegas, feeds Colorado River water to Southern California's MWD) often experience significant temperature differences in the summer between surface and subsurface waters. These differences result in a failure of waters at different depths to freely mix, further resulting in oxygen-poor environments that favor production of stinky and potentially noxious hydrogen sulfides, while increasing the ability of (often dangerous) minerals found in sediments at the bottom of the lake to dissolve into the water. As such, municipal providers drawing from deep bodies like Lake Mead or Lake Michigan can anticipate these changes and adjust their intake pipes seasonally if needed.

In contrast, planning for agricultural runoff after a massive rainstorm, inadvertent spills, urban discharges, and legal or illegal dumping proves more challenging. Some facilities, such as the Méry-Sur-Oise plant serving the downstream suburbs northwest of Paris (and thus the first to receive highly polluted urban wastewater), simply built physical barriers such as holding reservoirs to give themselves a needed temporal buffer between when a pollution event occurs and when the water reaches the treatment plant.[24] Philadelphia Water, which also draws on a slow moving river that runs through agricultural, industrial, and urban spaces before reaching the water treatment plant, similarly uses holding reservoirs. Chicago has no such option.

When one thinks of municipal water as a utility, something that needs to be safe so businesses can run and people don't get sick, the problem of off-tastes and smells falls to the bottom of the list of concerns; it would be *nice* to have good tasting water, but it's not necessary. However, once municipal water is conceptualized as a food, the consistent or intermittent presence of funny or unpleasant tasting water becomes a business and political problem: if people don't like consuming their municipal water, they may well go elsewhere. If someone regularly encounters new tastes or smells, they may lose trust in the utility, may be less likely to support infrastructure upgrades, and may mentally devalue their water supply, leading to increased waste. Most critically, for utilities managed by private companies or public-private partnerships, such as those that had come to define French water provisioning over the twentieth century, funny tasting or smelling water can undermine business relationships and undo contracts, all to the detriment of business growth.[25]

By the 1970s, municipal water providers for large urban areas such as Chicago, Philadelphia, Los Angeles, or Paris found themselves facing new twists on the taste and odor problems of earlier decades. Such problems generally emerge from four categories: from natural processes, from industrial chemicals, from treatment, and during storage and distribution. Some of the new problems remained closely linked to recently discarded or slowly accumulated industrial by-products in waterways. Others reflected the consequences of mid- to late twentieth-century innovations: elevated nitrate concentration from agricultural runoff (a hallmark consequence of the green revolution) increased levels of naturally occurring odor compounds produced by algae or microbial metabolism; treatment methods used to make water safe to drink unexpectedly interacted with compounds present in the water to create new flavor problems; and discharges of heated water from power plants supplying urban areas reduced oxygen levels in streams. Accumulated and novel pollutants, better detection techniques, and more stringent regulatory structures pushed water producers to diversify their methods of managing water's aesthetic qualities.[26] Off-tastes and smells in the late twentieth century necessitated new approaches.

The work of reconsidering approaches to tastes and smells included borrowing from other disciplines as explored in the previous chapter. It also opened doors to new collaborations with waterworkers from other countries, collaborations that traversed linguistic, cultural, and geographic barriers. In 1982 the International Association on Water Pollution Research (IAWPRC; established in 1965) hosted its first symposium entirely dedicated to exploring taste and odor problems, "Off-Flavours in the Aquatic Environment," in Finland. The organization officially connected scientists from more than twenty-one countries to collaboratively examine for the first time how to produce water in the face of quality and pollution challenges.[27] Water researchers recognized that working together, as well as drawing from the burgeoning field of sensory science, offered potential inroads to solving the problem of characterizing, identifying, and responding to smells and tastes in water.

This increased focus on taste came at the same time as a shift in opinion about the relationship between public and private enterprises in countries such as the United States, Britain, France, and Germany. Efforts to cut taxes, increase efficiency, and decrease overall costs of managing services including roads, waste, street parking, and municipal water production led to contracting out or selling off of assets previously owned and operated by public governments. Market competition, many economists suggested, could lead to better service than that provided by public entities.[28]

The rhetoric arguing that privatization would improve service proved especially powerful in France. Historically, the country has placed responsibility for municipal water supply on local towns (communes), while imposing national regulatory structures that govern the ways communes can operate. Communes can either choose to operate water production and distribution themselves or contract it out and delegate management to private or semiprivate companies.[29] This approach encourages public-private partnerships, creating a regulatory and management framework that offers significant earning potential for companies that promise to take the risks of upgrading infrastructure and collecting rents out of municipalities' hands. Many communes turned to private entities to help; by 1993, the Syndicat Professional des Distributeurs d'Eau, the professional water distribution association, reported 21,116 (57%) of France's communes were served by private drinking water supply companies.[30] The capital was no exception. In 1985 the city of Paris, under Jacques Chirac's mayoral direction, contracted water distribution to two private water management companies that had been operating in France for nearly a century: Lyonnaise des Eaux (which became part of Suez Environmental) for the left bank and the Compagnie Générale des Eaux (which became part of Veolia) for the right bank.[31] The decision to split the city's water management between two companies used the city's geographic layout to physically manifest the idea that market competition could improve water production.

While financial needs pushed city councils such as Paris's toward accepting privatization of their municipal water treatment, quality issues also acted as a pull. Competition for water provision contracts invited firms to expand their definition of water quality. Companies wishing to build their business portfolios could highlight in their bids to city councils and public-facing communication efforts the quality of water they could provide from a new perspective, that of the relationship between producer and tasting, smelling consumer.

The French move toward embracing private sector water supply contractors caught international attention. For example, the assistant editor of the journal for the AWWA, Ida M. Sayre, spent two weeks in 1986 touring treatment plants operated by the Compagnie Générale des Eaux. "Competition among private sector water supply contractors seems to stimulate pride in the appearance of treatment plants as well as their efficacy," Sayre noted in her report, pointing out that competition also "mandates providing quality service at a reasonable price." Sayre lauded the trickle-down of expertise, resources, and know-how from the parent company to small municipalities, pointing out the rapid increase in access to chlorinated water for some small communes as a result of privatization.[32] For the larger towns Sayre visited, privatization's benefits extended beyond offering citizens access to water essentially guaranteed to be potable for the first time. Privatization also, Sayre pointed out, offered the combined technical know-how capable of controlling significant pollution challenges caused by agricultural intensification, industrial pollutants, and accidental spillage. The Méry-sur-Oise plant, which supplied water to the north Paris region, dealt with water containing heavy organic and inorganic compounds, accidental pollution from shipping accidents, leakage from industrial waste lagoons, and failures of upstream wastewater treatment plants. Privatization's reach, Sayre pointed out, justified extended research to develop and verify new processes for removing novel pollutants such as chromium as well as more widespread ones such as high nitrite levels. In summary, Sayre saw expertise as circulating freely throughout an enterprise's network, synergistically enabling innovation at scale.[33]

Lyonnaise, like the Compagnie Générale des Eaux, used its expanding research capacity to argue for its ability to manage water. In 1981, under newly appointed president Jérôme Monod, Lyonnaise inaugurated a new research center, the Centre International de Recherche sur L'Eau et l'Environnement (CIRSEE). The shiny new center housed Lyonnaise's research arm, a built manifestation of the company's move toward expanding the business onto the international stage through technical dominance over competitors. François Fiessinger, CIRSEE's director, understood the building as a knowledge-production hub at the center of an ever-expanding national and international network of operations.[34] Fiessinger's work as director proved successful, both for his career and for the company. He helped judge an eight-city taste test between city waters in Washington state in 1985, the same year Paris awarded Lyonnaise the contract to manage water on the left bank; the Association of Environmental Engineering Professors selected Fiessinger as the distinguished lecturer for 1987; and by the 1990s, Lyonnaise obtained contracts in cities around the world through a series of local-based partnerships.[35] Lyonnaise's expertise circulated through its increasingly international network, helping win contracts in Buenos Aires, Mexico City, Cartagena, Sydney, Chengdu, Johannesburg, Jakarta, Manila, La Paz, Budapest, Maribor, Casablanca, Medan, Santiago, and Atlanta, as well as cities within the United Kingdom and

Spain.[36] For Lyonnaise, making better-tasting water emerged as an inflection point in the firm's larger, strategic goal to gain municipal water contracts around the world, a strategic goal grounded in the idea that through research Lyonnaise could manage water quality in a way that kept the future sensing body of not-yet-obtained customers content with their water.

## BREAKING ISOLATIONIST BARRIERS

To build Lyonnaise's ability to make better-tasting water, Fiessinger stepped away from the isolationist approach held by its major competitor, Veolia, and built an international network of expertise. Fiessinger started by recruiting scientific experts in the major fields for water treatment from the membership of the AWWA in 1984. Collaborators included Charles O'Melia from Johns Hopkins University (with expertise in coagulation); Vernon Snoeyink from the University of Illinois at Urbana Champaign (with expertise in activated carbon); Joel Belfort from Rensselaer Polytechnic Institute (with expertise in membrane technology); and in the coalescing field of taste and odor chemistry, Irwin (Mel) Suffet from Drexel University. By actively reaching out to water researchers in the United States, Fiessinger expanded the technological expertise of employees at Lyonnaise's CIRSEE. More critically, from the business perspective of a company trying to increase its footprint, these international collaborations offered to help transform Lyonnaise's technical capabilities to quickly test and respond to problems identified in Lyonnaise's growing network of contracts with municipalities inside and outside of France.

Prevention of off-flavors played a key role in the collaboration between Lyonnaise's French and US-based water researchers. Auguste Bruchet, who in the early 1980s was working as a technician at France's Lyonnaise des Eaux, recalled that in the early stages of the collaboration, "each time that someone brought in a water that had an odor—in general we talk of tastes but most of the time we are actually referring to odors—we began to identify some dozen components but we never knew which were causing the problem."[37] Like the chemists using GC in the early 1960s, Bruchet and colleagues quickly found themselves concluding, "'We've identified some compounds, they are all below the odor threshold, but perhaps the synergistic effects between these components can explain the observed odor.' In other words, our diagnostic capacities were zero."[38]

Bruchet paints a grim picture of waterworkers' ability to actually characterize unwanted tastes and smells in a way that could scientifically guide treatment approaches, in contrast to the hopeful rhetoric engineers and chemists in the United States had put forward in the 1930s and 1940s. Lyonnaise and partners' inability to successfully link compounds identified with treatment options started to change, Bruchet recalled, in 1984. As technical know-how, students, and research questions traveled between Lyonnaise's research facilities outside of Paris and Mel Suffet's lab at Drexel in Philadelphia, new opportunities for explaining the cause of particular odors emerged—but those opportunities depended on the ability of researchers from different locations to effectively communicate across linguistic barriers shaped by culturally informed ways of talking about tastes and smells.

To ensure that researchers at each site could compare results, the team adopted FPA, the consensus flavor description technique developed in the 1950s by the consulting firm ADL. FPA quickly came to define what I think of as the "Stinky Influent" school of water treatment that emerged through the collaboration between the Philadelphia Water Department, the Philadelphia Suburban Water Company, and Lyonnaise des Eaux. Getting all three water treatment utilities to consistently use FPA depended on circulating expertise among the different locations. Under the direction of Suffet, then graduate student Djanette Khiari led the majority of training. Khiari, who later became a research manager at the Water Research Foundation, gained expertise in FPA while interning at the Philadelphia Water Department under Irene Taylor, its coordinator of FPA.[39] Khiari subsequently became the "trainer in chief" for Suffet's lab as well as workers at Lyonnaise.[40] "The French were very serious about [taste and odor]" Khiari recalled. "They have little tolerance for chlorinous odors."[41] What started out as a project with two leased mobile labs quickly expanded into a multiyear effort to teach people how to verbalize odor, arrive at consensus, and develop shared language.

Although *Standard Methods for the Examination of Water and Sewage* had not yet adopted FPA, the international collaborators found using it facilitated their efforts to link off-flavors with their specific molecular causes. For example, when researchers at Lyonnaise began preconcentrating water samples before GC-mass spectroscopy analysis, the increasingly sensitive detection technique allowed them to, as Bruchet put it, "identify certain compounds [in the water] that had only been identified for the first time within the previous ten years such as geosmin and methylisoborneol, the two principal causes of earthy musty odors worldwide."[42] In turn, this preconcentration method facilitated linking the presence of a strong medicinal odor in water with chlorination by-products in 1986.[43]

A molecularized understanding of taste and odor as capable of activating consumers' aesthetic perceptions shaped efforts to link off-flavors with their specific chemical, but not necessarily environmental, causes. Success with FPA suggested that the holy grail appeared within reach where the molecular cause of an odor could be identified and used to navigate its potential future impact on future tasting bodies. One simply needed an appropriate library of reference standards to link with the wealth of descriptive terms generated over the century by water tasters; those standards and descriptive terms could in turn guide treatment to reduce the concentration of offending molecules to below the threshold of perception before they entered the distribution system.[44] Bruchet recalled, "We began to be able to identify certain compounds [in the water] that had only been identified for the first time within the previous ten years such as geosmin and methylisoborneol, the two principal causes of earthy musty odors worldwide."[45] Despite the excitement of linking molecules with sensation, it remained difficult to find easily shareable descriptive terms that crossed language barriers. Words alone were insufficient.

## LINGUISTIC GAMES: DO YOU SMELL STINK BUGS OR LETTUCE?

The French-US research collaboration highlighted the critical need for a language capable of crossing international borders. Without training, people struggle to describe odors. Scratch and sniff tests, plus repeated exposure to standards, helped trainees gain the ability to name odors. While certain descriptors were easily shared, others carried with them specific cultural references that did not easily translate. Khiari recalled that linguistic discrepancies didn't significantly affect intra-lab conversations. They could, however, trip up conversations between laboratories. Scientists from the United States routinely described hexanal—a common standard used in training panelists—as lettuce. The French, in contrast, called it stink bug. "That's cultural. That's part of, it's the way, you know, you grow up," Khiari remembered.[46]

For small groups working closely together, linking descriptors like lettuce and stink bug was simply a matter of conversation and a few shared laughs. In contrast, adapting sensory techniques for use by an international cohort of practitioners tasked with identifying and treating tastes and smells required developing a shared language capable of functioning between separate labs in separate places. Members of the IAWPRC responded by forming a study group focused on tastes and odors. Per-Edvin Persson, then a research scientist at the University of Helsinki, recalled that as chair, "One of the tasks I pushed in the Specialist Group was the standardization of sensory techniques and, in particular, the descriptive terms of different odours [*sic*]."[47] The Study

Group on Tastes and Odours held its first meeting in Paris in 1985, hosted by the city, in an effort to begin standardizing descriptions.[48] The group came up with forty descriptive terms for the flavors found in natural waters. "Three descriptors were easily agreed upon," Persson noted. "Earthy for geosmin, musty for 2-methylisoborneol, and moldy cork for 2,4,6-trichloroanisole. The moldy cork was a French contribution (bouchon moisi)."[49] The other thirty-seven descriptors proved more challenging—although producers recognized the identified molecules as causing off-tastes and smells, they could not agree on linguistic descriptors to help communication.

But why should language matter in a world where scientists could use instruments to detect the molecules responsible for flavor? The reality, as Bruchet pointed out, was (and remains) that humans are often more sensitive detectors than instruments.[50] "Classification of odors that may be detected in drinking water is desirable and, in fact necessary," argued Suffet and Lyonnaise employee Joël Mallevialle in their 1987 publication *Identification and Treatment of Tastes and Odors in Drinking Water*.[51] The manual, produced through the AWWA Research Foundation (now known as the Water Research Foundation) (Suffet)–Lyonnaise (Mallevialle) collaboration and funded by the French government, specifically sought to help waterworkers mitigate public rejection of municipal water due to aesthetic concerns by disseminating FPA throughout the industry.[52] "Is the public rejecting perfectly safe water due to aesthetic concerns while choosing to drink palatable waters that are not safe?" Mallevialle and Suffet wondered, sternly noting, "Palatable waters are not always potable."[53] Despite the descriptions given to the press of Parisian water tasters successfully finding and suppressing off-flavors before they reached consumers, dealing with the wide range of gustatory contaminants was nearly impossible. Water tasters, even those trained in FPA, struggled to consistently articulate descriptors that matched sensation.

## FROM WORDS TO WHEELS

Waterworkers, represented by collaborators from AWWA Research Foundation, Lyonnaise, and IAWPRC's study group on tastes and odors, wanted a road map that cut across their diverse geographical and cultural landscapes to help them identify the best treatment approaches. To address this desire, in the mid-1980s waterworkers turned to two recently developed tools: a flavor wheel developed by the beer industry, now termed the Beer Flavor Wheel, and one developed by the Sensory Evaluation Sub-Committee of the American Society for Enology and Viticulture, now known as the Wine Aroma Wheel.[54] Both acted as "intersubjectivity engines": tools that help users to "reliably assign stable descriptors to odors and tastes" and in the process result in the formation of taste communities.[55] Both discarded subjective terms related to liking (hedonic values) in favor of terms that could be readily linked with standards. The beer industry's wheel highlighted odors as well as sweetness, sourness, bitterness, saltiness, and mouthfeel. In contrast, the Wine Aroma Wheel (see figure 4) focused only on aromas, excluding mouthfeel and visual cues; this move stepped away from the "imprecise and non-objective" tasting techniques of remarking on color and balance commonly used by wine critics or sommeliers.[56] The Wine Aroma Wheel aimed to facilitate identification of the tasty and aromatic details of a wine's terroir (taste of place) for experts and novices, while the Beer Flavor Wheel was aimed primarily at an expert audience.[57] Sensation and words came together into a simple-to-use communication device.

These graphical representations of the flavor profiles of beer and wine caught the attention of water scientists working to manage off-flavors, not because of the tools' potential for disrupting the divide between expert and novice, but because of their potential for facilitating collaboration across cultural barriers. Both the Study Group on

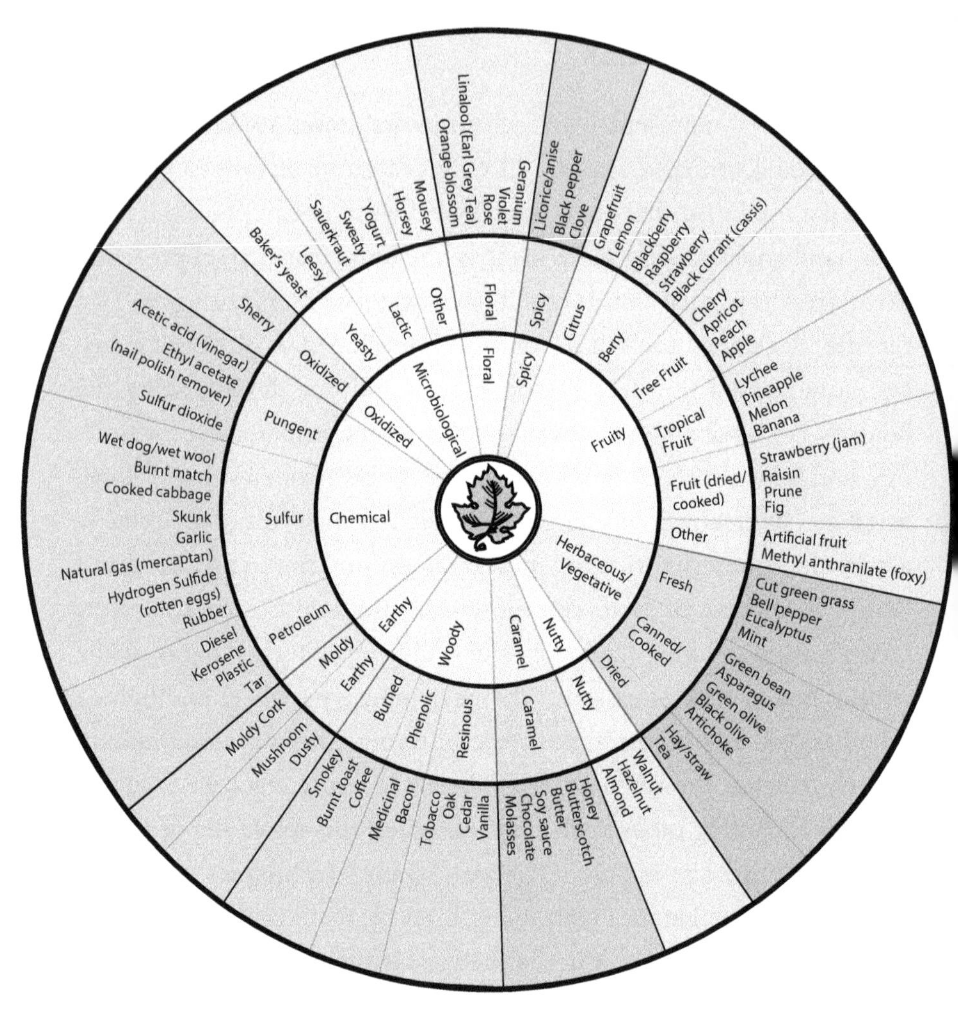


**Figure 4.** Wine Aroma Wheel. Includes classification categories in inner circle, subdivided into two levels of increasingly descriptive specificity. *Source:* A. C. Noble, copyright 1990 (www.winearoma wheel.com). Reprinted with permission.

Tastes and Odours (IAWPRC) and Suffet and Mallevialle adapted the idea of the flavor wheel, developing graphical, wheel-like representations to accompany their lists of standardized terms. Suffet jokingly described this as stealing the food and wine industry's work: "[We saw] a good idea and ran with it in terms of what we were interested in."[58] Similarly, Persson described it as a process of mimicking.[59] Imitation allowed something new to emerge: a technology to help transform the work of municipal water production in multiple locations into something that consumers could more easily ignore.

The wheel produced by Persson's IAWPRC group and the wheel proposed by Suffet and Mallevialle differed in little and large ways from the beer wheel and the Wine Aroma Wheel. For example, although the IAWPRC wheel mimicked the form of the Wine Aroma Wheel—a circle with major categories at the center and concentric rings indicating increasingly specific descriptors—many of the descriptors (DMS, aliphatic solvent, plasticizer) reflected specialized knowledge not available to the average person or to a newly hired engineer trying to navigate water's unexpected gustatory qualities. The descriptors in the IAWPRC wheel marked the realm of technical expertise and required knowledge gained through training, but "lacked [descriptive terms] for many of the odors associated with treated drinking waters and groundwaters."[60] Suffet and Mallevialle's proposed Flavor Wheel for Raw and Treated Drinking Waters was built on wheels from the beer and wine industry as well as the IAWPRC. Instead of the concentric circles used by IAWPRC, the wheel uses thirteen pie slices of more or less equal size similar to that of the Beer Flavor Wheel (see figure 5). Subsequent iterations in 1991, 1994, and 2016 of the flavor wheel, now renamed the Drinking Water Taste and Odor Wheel, look increasingly like a mixture of the Beer Flavor Wheel and the Wine Aroma Wheel; they use words easily associated with substances encountered during everyday life, adding on associated molecular reference standards to the outside of the wheel.[61] For water workers unfamiliar with FPA, Suffet and Mallevialle's revised wheel facilitated linking familiar off odors with their potential chemical causes (see figure 6).

If you were to pit the adjectives for the Wine Aroma Wheel against those of the water wheels, the Wine Aroma Wheel would win every time. An overwhelming number of undesirable flavors characterize both water wheels. The IAWPRC wheel offers up descriptions of water

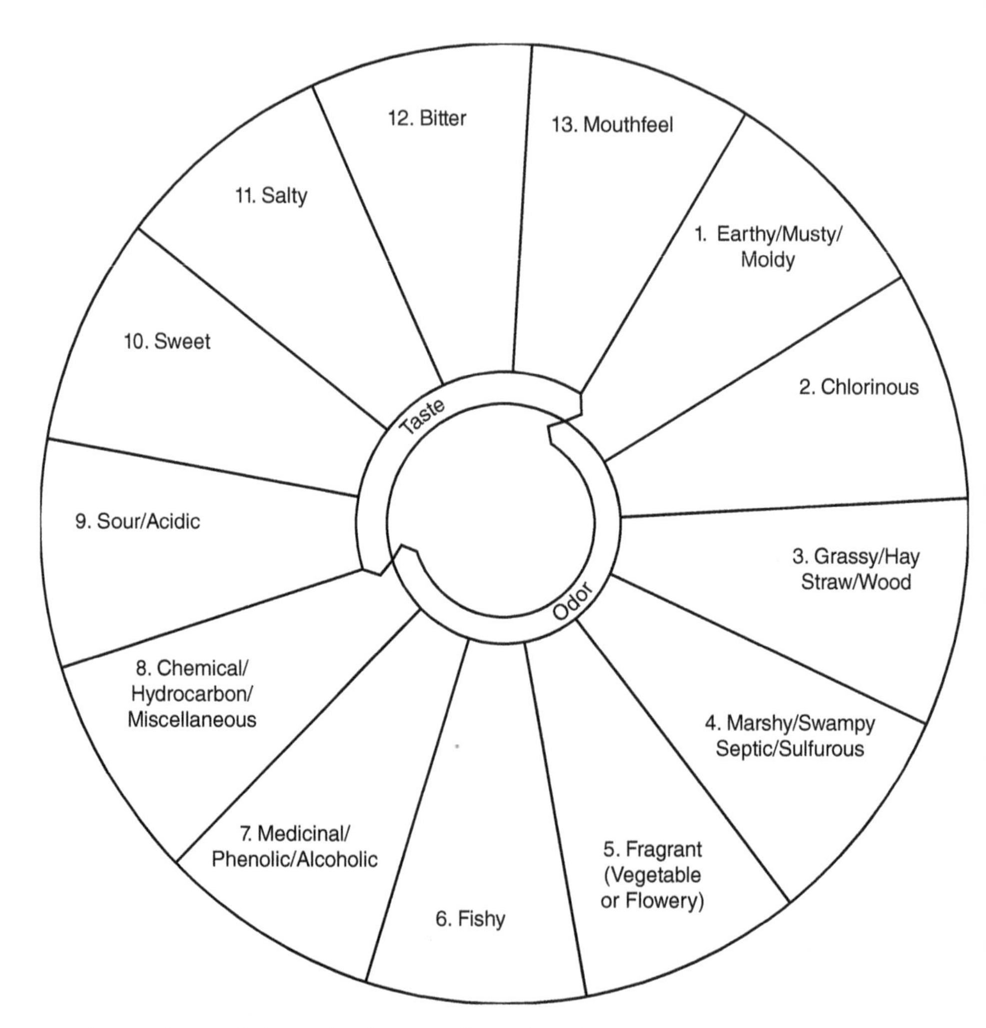


**Figure 5.** Proposed Flavor Wheel for Raw and Treated Drinking Waters. Contains eight odor categories, four taste categories, and one mouth-feel category. Unlike later wheels, it presented potential reference compounds in text with aim of jumpstarting "rapid standardization of the taste-and-odor reference materials in the drinking water industry" (p. 109). Reprinted with permission from *Identification and Treatment of Tastes and Odors in Drinking Water*, 1st Edition, Copyright 1987 the American Water Works Association. All rights reserved.

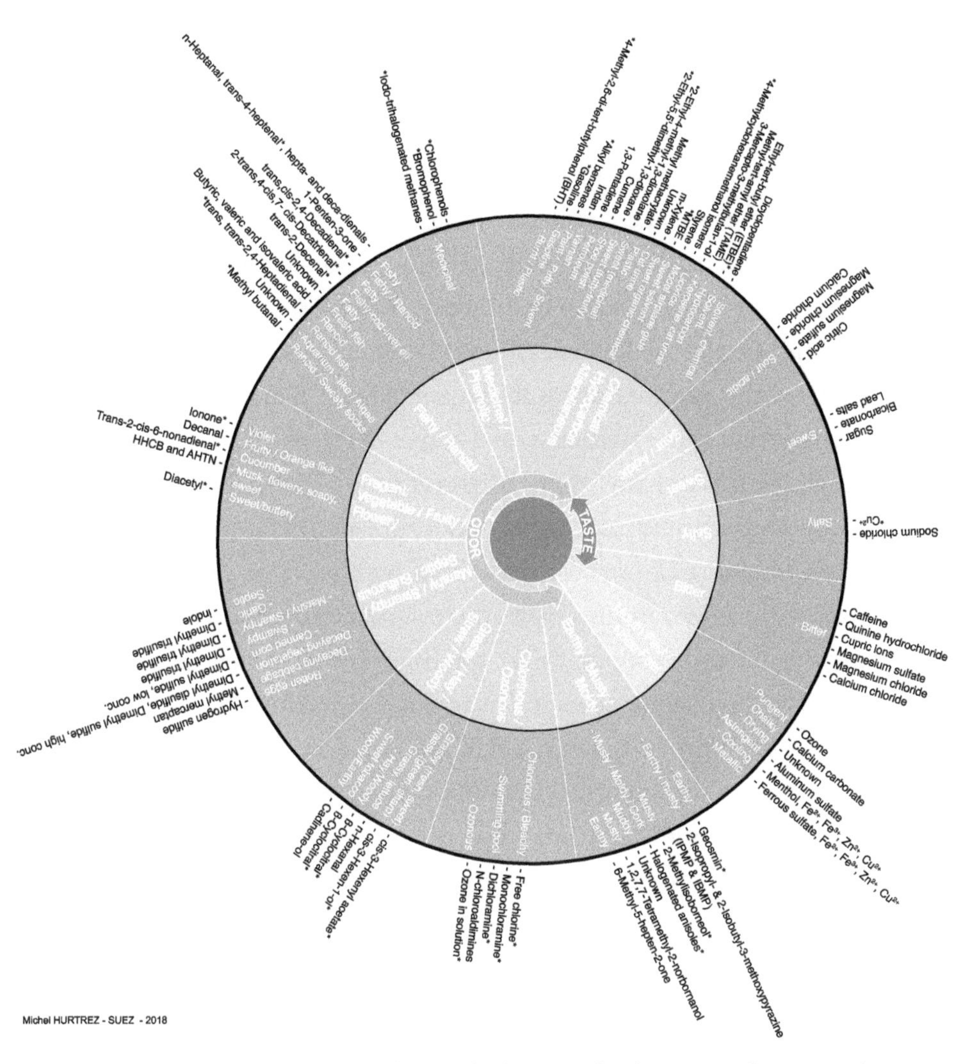

**Figure 6.** 2016 Drinking Water Taste and Odor Wheel. Reprinted with permission from *Taste and Odour in Source and Drinking Water: Causes, Controls, and Consequences*, Copyright 2019 IWA Publishing. All rights reserved. Contains eight odor categories, four taste categories, and one mouth-feel/nose-feel category in inner circle, surrounded by outer circle that includes common taste and odor terms developed by trained sensory panels. Outer rim of wheel presents specific chemicals as reference standards for odors. Authors marked "known" standards with *; others are "representative" compounds.

as chemical, medicinal/disinfectant, natural unpleasant, earthy, and natural pleasant. Essentially more than three-fourths of the aromas identified for untreated waters the IAWPRC group examined are for off-flavors. The majority of the groups in Suffet and Mallevialle's original wheel characterize disagreeable odors: Earthy/Musty/Moldy, Chlorinous, Grassy/Hay/Straw/Wood, Marshy/Swampy/Septic/Sewage, Fragrant (Vegetable or Flowery), Fishy, Medicinal/Phenolic/Antiseptic, Chemical/Hydrocarbon/Miscellaneous. The descriptors present on the original and subsequent wheels mark the mundane reality that raw water carries with it traces of the natural and human-made places a water has traveled through; for example, sweetness, fishiness, and earthy/musty/moldy odors are all linked with open reservoirs and biological activity; bitterness with elevated pH caused by corrosion of cement linings or metal leaching; mouthfeel or nosefeel with galvanized or copper pipes, corrosion, or even new plumbing; marshy/swampy/septic/sewage with poorly circulated or maintained reservoirs or improperly maintained hot water systems; and so on.[62] The descriptors also highlight the political and cultural ways in which water in the twentieth century came to be valued: as a resource to be held in reservoirs, to be piped underground and through buildings, or even to be used as a dumping ground.

## UNDOING TERROIR

One is struck by the disagreeableness of the olfactory landscape described by the water aroma wheel: Who wishes to carefully savor a water that smells of hydrocarbons, sewer, or mold? The descriptors highlight the dual challenge of creating unremarkable water, a challenge split between naturally occurring microorganisms and human-added inputs. All can contribute to production of the described odors. Research in 1984 comparing common odors of both influent and effluent waters at the Philadelphia Suburban Water Company (Suburban), the Philadelphia Water Department, and Lyonnaise des Eaux (Lyonnaise) found similar odors at all three locations, although at different frequencies of occurrence. Influent from Suburban was described as "sewage, creeky, musty"; Philadelphia Water Department's influent was "decaying vegetation, septic, vegetation, earthy, musty, fishy"; and Lyonnaise's exhibited "muddy, fishy, musty, septic" characteristics.[63] The fact that all locations drew from river waters certainly contributed to the similarity between raw water at these different locations. But so did pollution from an increasingly similar set of activities that cut across geographical and cultural distance: a boom in discards from urbanization and industrial agriculture. In contrast to wine, waterworkers wanted tools that could help undo the markers of terroir found in the water coming into their treatment facilities.

Surface waters are especially susceptible to contamination by purposeful or careless behavior. Both Suburban and Lyonnaise draw their supply from surface and ground waters; solely surface waters supply the Philadelphia Water Department. All of the researched watersheds are located in historically industrial zones and include significant agricultural activity; both industrial and agricultural discharges have been linked to increased growth of odor-causing microorganisms, as has drought.[64] The similarity in challenges faced by Lyonnaise, Philadelphia Water, and Suburban reflect a mid- to late twentieth-century convergence of industrialization, urbanization, and pollution.[65] This convergence shows up in the textual and graphical interface of the water wheels, revealing raw water as a material marked by human activity and carelessness. In fact, the water aroma wheels reflect a material best avoided until the wizards of water treatment render it potable and presentable, a far cry from the images of pristine water pulled from untouched aquifers portrayed by many bottled water purveyors.

The success of the collaborative research partnerships embodied in IAWPRC's working groups, and that of the AWWA with Lyonnaise, created another inflection point in the work of managing the tastes and smells of water. Water wheels collapse time and geography. In building the water wheels, these entities created their own road maps, unique to the material they worked with. The descriptors point to sensation. But they also point to the *molecular* causes of certain sensations. In short, the descriptors link an olfactory cue with a class of molecular substances and possible causes, and in the process direct waterworkers toward specific treatment approaches. The wheels codified a way for municipal water utilities around the globe, many of which are part of the Lyonnaise (now Suez) network, to identify and treat off-flavors. Improved technical and analytical efficiency allowed water producers to detect ever smaller amounts of molecular contaminants: over a period of fifteen or so years, detection improved to between 20 and 50 picograms/liter. Bruchet, who is now retired, has estimated that his team could identify the cause of off-flavors in water 70 percent of the time.[66] Identification, in general, is quickly followed by treatment to remove the offending molecules, a practice consumers remain blithely unaware of until it fails. In short, success in understanding and treating the industrially inflected tastes and smells of raw water allows polluting activities to continue without any outcry from consumers and without requiring cities to search for new water sources; in contrast, failures in treating these tastes and smells can push consumers into polluting behaviors such as the purchase of bottled water. No matter this double bind, the larger work of removing the human-caused sources of unwanted tastes and smells remains outside of most municipal water producers' political reach.

In 2005 Lyonnaise discontinued its research program on off-flavors and odors in water. Bruchet attributed the program's closure to a lack of new analytical techniques.[67] Instead, the company turned toward characterizing undesirable odors from waste, a move that reflected the company's shifting business strategy. For now, faced with aging infrastructure and austerity measures, municipal water has reached a "good enough" state of technical triumph over water. Although obdurately imperfect, the water wheel, and the accompanying technologies that helped make it possible, set in motion a new era of taming off-tastes and odors characterized by actively encouraging waterworkers to perform sensory labor and discouraging consumers' sensory labor.

## INEFFABLE AESTHETICS

Tastes and smells do not just exist. They are artifacts of how living things interact with nonliving things, of summer heat and water stratification, of agricultural runoff and algae blooms, of minerals dissolving or pipes releasing chemicals into a water supply. By trying to make embodied experiences accessible to others, the collaborative efforts of waterworkers in the 1980s and 1990s between Europe and North America added a new inflection in the twentieth-century journey of industrializing water by allowing workers to manage the sensory experiences of the future sensing body. Located in different times and spaces from where the work of managing taste is done, the future sensing body demands treatment of the industrially inflected aquatic environment. A combination of techniques, from water wheels to FPA to closed-loop stripping, GC, and mass spectroscopy, combined to make the musty, swampy, or whatever-y sensations found in raw water into objects that *can* be known and treated (assuming enough resources). This blending of techniques goes beyond the treatment plant. As waterworkers sought to improve their ability to manage perceptible markers of quality, they also worked to manage the uncertainty of raw water, a stopgap measure for navigating the human and environmental systems that make raw water such an uncertain ingredient. By learning to manage quality in such a way that the future sensing body became an aggregate body *unaware* of raw waters' tastes and smells, waterworkers also became the arbiters of what flavors could circulate. Indeed, this industrial logic of flavor management, made possible by the shift to understanding tastes and smells as molecular *and* by the increased ability to communicate across different water treatment plants about challenges, set the stage for the twenty-first century move to promote the use of treated wastewater for DPR. If we know water well enough, we can make it clean enough to drink without ever introducing the "middle manager" of the water cycle.

It may be easy to recognize the cultural politics of who has Taste and who does not when sitting around a table tasting a glass of wine. It is especially easy when that tasting is done in a way that marks class and knowledge: holding the glass up to the light, swirling, sniffing, sucking a small sip in, and then gargling before finally announcing whether you would buy the bottle, or outlining the flavors found in the glass. In contrast, the sensory labor employed in understanding, identifying, and talking about taste and smell by waterworkers in the United States and France has, through processes of imitating and changing the sensory labor done by wine tasters, made it harder to identify the various politics at play in the making of water. The next chapter examines what happens when tasting practices become sensory labor, work moved from the back-of-the house arena of water research labs and treatment plants to streets and boardrooms through explicitly teaching people from all walks of life to attentively taste water in a way that produces value for others.

# 4

# THEATERS OF TASTE FROM THE BOARDROOM TO THE STREET

IN 2016, amid extensive media coverage, the Fine Water Society held its first International Fine Waters Tasting Competition in Guangzhou, China. Video reports showed a group of smartly dressed male water sommeliers carefully nosing, swishing, and spitting still and sparkling waters from around the world.[1] Water-tasting competitions like this have existed for a few decades (the Berkeley Spring International Water Tasting, which includes municipal and bottled waters, celebrated its thirty-third event in 2023). Established in 2008 by Michael Mascha and Martin Riese, the Fine Water Society aims to "enhance the visibility of Fine Water and convey the idea that water is not just water but a natural product with terroir and unique characteristics."[2] The society offers a water service certification, as well as the opportunity to become a certified water sommelier. It additionally promotes adoption of fine waters—what Mascha defines as the "world's most distinctive bottled waters."[3] As the judges at the Fine Waters

Tasting Competition nosed, swished, swallowed, and rated each water on a scale up to 100, they used multiple senses to evaluate the waters. Which aspects each judge valued on the standardized scales used determined which waters "walked" away with which prizes.

The rise of the water sommelier and the practices of conceptualizing, evaluating, and drinking waters like fine wines is a recent turn (although one preceded by Paris's municipal water tasters). In fact, teaching people to consciously taste is a hallmark of the wine world: to drink wine "well" you need to learn the language and conventions surrounding the beverage. Someone who wants to gain experience can attend an evening training with a sommelier, take a whole series of courses, or simply purchase a copy of the Wine Flavor Wheel online and set about training themselves. Formal or informal, all of these trainings educate tasters to pay attention to some things and not others.

I don't drink wine. Nevertheless, I have participated in two wine tasting trainings as part of classes on French food culture; both introduced me to a basic sequence of steps to help one pay attention: look, smell, taste, and think. I now know to swirl my glass and hold it up to the light, to observe the "legs" as the wine drains back down the sides of the glass—a visual shortcut to evaluating a wine's sugar or alcohol content—before sticking my nose in the glass and sniffing. I then compare notes with colleagues who taste to see how I did. I am similarly trained in tasting Comté cheese. Every time I eat a slice, I reproduce the training I received while studying abroad by observing the color, breaking the piece of cheese open, watching how it breaks, and then sniffing the piece before placing it in my mouth. As I chew, I observe the texture (smooth? grainy?), then suck in a small breath through my teeth before breathing out of my nose to capture the cheese's aromas. I pause to notice the sweetness, saltiness, and other tastes, before finally swallowing (I like cheese too much to ever spit it out).[4] Both trainings changed how I approach foods. The cherry, jammy wines I sniffed and nutty, grassy cheeses I tasted invited me to carefully think about what my mouth and nose were communicating to me about the regions these foods came from and their production processes. These trainings did more: they taught me to taste with the goal of understanding, something the French term *dégustation*. Dégustation is a tasting technique that invites eaters to use the tasting experience to place food within a larger system. For example, when I look to see whether the Comté I am eating is a pale white or darker yellow color I am looking to understand whether the milk came from cows fed winter hay or summer grass. Both are feeding decisions shaped by the natural environment and local regulations around Comté's production. Being taught to taste reoriented me toward new ways of thinking about and relating to the product consumed.[5]

My education in observing wine and tasting Comté altered how I use my eyes, mouth, and nose when I encounter foods. Taste education aims to shift the register of its students' sensory labor from unconscious attunement to the environment into conscious practices of paying attention. Willing engagement in taste education can transform how one tastes, Geneviève Teil and Antoine Hennion suggest, by activating people's ability to "perform" tasting by using appropriate techniques or props (e.g., the right glass for the drink) and then reflecting on their experience.[6] For Teil and Hennion, the performance emerges as one does the activity, drawing on "aesthetics and physical and mental conditioning," to name, identify, and learn.[7]

People are really familiar with the tastes, smells, and textures of the water they regularly use. As someone repeatedly engages with their environment they become a "measuring instrument," attuned to nuances that someone new to the area may miss.[8] That recurrent exposure results in the development of everyday expertise, the tacit, embodied knowledge about what is or is not "normal."[9] In other cases, like the training in FPA conducted at the MWD of Southern California or Lyonnaise des Eaux, or the water sommelier certification, taste education trains someone to notice certain things and ignore others.

Reflexivity, naming, and learning shaped the categories the water tasters in China paid attention to at the International Fine Waters Tasting Competition. A series of socially trained gestures defined *how* each judge paid attention. As the judges tasted the waters and rated them, they followed a similar pattern to waterworkers trained in FPA. Unlike waterworkers, however, the judges were not tasting like potentially interchangeable experts. Rather, they each had been trained to taste like a connoisseur, to perform their aesthetic expertise for an audience. For the waters submitted to the competition, the outcome could significantly boost their value. The judges' evaluations sought to capture each water's terroir in a way that could then be used to design menus and pairings and encourage future consumers to learn to appreciate fine waters like others do fine wines.

However, as this chapter demonstrates, learning to attend to mineral water's terroir means pretending that what we get in the bottle is entirely natural. Attending to terroir is a mode of paying attention that also creates ignorance. It purposely shapes the relationship between producer and consumer to only highlight one part of the production story. It is true that mineral waters taste of the earth they have traveled through. Nonetheless, attending to a bottled water's terroir is a move against the standardization of taste critical to municipal water producers' efforts to create a product that doesn't call attention to itself except when it is out of the ordinary.[10] New practices of training the public in tasting water that emerged in the early twenty-first century are an inflection point in the management of how urban consumers understand the quality of their tap water.

## PERFORMING SENSORY LABOR

Bottled water producers found 2008's economic downturn difficult. Sales dropped globally. Analysts pointed to the credit crunch for partially causing the weak market for bottled water. But they also blamed "highly damaging media exposure" brought about by campaigns promoting tap water that critiqued the hidden environmental costs of bottled water consumption.[11] As sales numbers continued to decline, producers of bottled mineral waters looked for new ways to engage bottled water consumers and new methods to demonstrate the value their waters provided.

Nestlé Waters responded to the 2008 downturn by working to redefine how people thought about Nestlé's waters. The company thought that promoting the unique individuality of each bottled water offering in its water portfolio could be key to revitalizing sales. This insight, ironically, came from efforts to expand the portfolio's most standardized product, Pure Life, into new markets. Like Dasani and Aquafina, Pure Life is a still, bottled water that contains remineralized treated water from local sources (municipal or well water). "At the beginning we worked at Nestlé's research center, based in Switzerland, with experts on a variety of subjects and attempted to find a relatively neutral taste," a former member of the Nestlé communications team, Elle, told me.[12] At first the team had imagined they could determine and define the perfect everyday water using the molecular logics of flavor: they could take everything out via techniques like reverse osmosis (RO), then add back in the "ideal" mixture of minerals. Blind consumer tastings showed otherwise. While consumers in one country liked the standardized version of Pure Life, consumers elsewhere rejected it. Given the business charge to make Pure Life a stronger market player than local water offerings, be they bottled or from the tap, Nestlé reformulated Pure Life into a variable water, one that reflected regional taste variations. In some cases that meant mimicking mineral water, and in others, offering an almost mineral-free water. Even bottled and standardized, Pure Life's success depended on reflecting place.

Building on the "discovery" that local tastes mattered for a product as anodyne as Pure Life, the communications team embraced a European, terroir-centered approach to building out the marketing campaign for the Nestlé water portfolio. The team started by hiring outside experts to define the tastes of their water portfolio; the group included tea master Yu Hui Tseng; a Franco-American sommelier, Jeremy Moreau; and Andreas Larsson, the world's best sommelier (2007). The group was rounded out by George Lepré, director of the Professional Studies Program at the Academy of Wine in Paris, and Ghislain de Marsily, a hydrogeologist known for his work on subterranean water contamination and a member of the prestigious French Académie des Sciences.[13] They spent three days together at Nestlé's Issy-Les-Molineaux headquarters tasting a variety of waters from around the world, with the understanding that their findings and insights would eventually be distilled for public consumption.[14]

Rather than drawing entirely on the rules of connoisseurship that had defined wine tasting for much of the nineteenth and early twentieth centuries, the tasting employed a hybrid model that combined practices of connoisseurship with sensory science laboratory practices. Different waters were presented blind, without identifying information. By blinding the identity of the sample under examination, the Nestlé team aimed to remove as much subjectivity as possible from their human measuring instruments.[15] In bringing together a tea master, sommeliers, educators, and researchers, the communications team reflected the wine world's understanding that people with trained expertise were the best fit for doing the sensory labor of tasting and defining a beverage. In the tasting room, the tasters performed how recognizing terroir depends on training.

## THE COMPLICATED NATURE OF TASTING NOTHING

Water and wine, despite the links to soil highlighted through the framing of water as having terroir, are very different materials. Outside of the odorous compounds produced by pollution or by microbiological action, most of water's flavor comes from dissolved minerals. Aroma perception largely occurs when volatile or semivolatile compounds enter the mouth, nose, or the passageway connecting the two.[16] Procedures for tasting wine, with its much larger percentage composition of smell-able compounds, do not necessarily translate into tasting water. As one of the experts present at the tasting, Gilles told me that tasting water is a lot more subtle than tasting wine due to the minimal number of aromatic compounds in water. "You really have to get in it and go look in the far corners of your glass to find something to say about the actual aromatic components," he pointed out. "A lot of tasting water is assessing it on the palate, the balance of it, you know the acidity, etc. Whereas with a spirit, 90 percent of the experience is nosing and the palate is just kind of confirmation of information of what you've nosed previously. Tasting water is the hardest thing I've ever done. It's brutal."[17]

Elle pointed out that for the employees taking part in the tasting, the differences between waters were almost impossible to identify. The experts were much more sensitive: "During the testing session we included a little trap; we took tap water and put it in a bottle and placed it among all the other samples," Elle noted. "Honestly, Madame Tsing almost vomited [when she tasted it], remarking, 'Oh, this is horrible, what is this water?'"[18] Gilles reported, "The most interesting for me was realizing that mineral water, well, we couldn't really explain it at first when we were doing the blind tasting, but we always kind of hovered around the same words; like certain waters tasted dead, and others tasted alive. We realized that blind tasting them 99% of the time we were able to tell a mineral water from a tap water or a spring water."[19] Tasting was seeing truly, with an estimated, albeit debatable, 1 percent error: the tongue and nose undid the trap and revealed tap water's chemically treated presence to the expert.

The moment the experts' mouths and noses revealed the *piège*/trap enacted something important. In setting up the tasting by blinding the samples, the organizers of the tasting event posited a specific molecular reality that linked the experts' mouths and noses with the biogeographical origins of the waters tested. By including a *piège*/trap in the form of municipal water, the organizers similarly posited a specific, perceptible difference between "natural" mineral water and "processed" tap water. The tasting validated the effort, as Elle put it, "to truly show that mineral water has something equal in contrast to chemically treated tap water" by mobilizing sensation to charismatically produce knowledge—not just of the differences between different bottled mineral waters, but also of the difference between bottled and tap water.[20] One might see this as an odd way to demonstrate that bottled water's most obvious competitor tasted different. After all, the audience was small. Only people employed in some fashion or another by Nestlé participated. Knowledge about the test's details, including the little trap, never entered the marketing tools generated by the test. But the test did more than validate bottled waters as distinct from municipal waters; it reoriented—or further oriented—those promoting mineral waters toward the idea that flavor and texture offered their own truths that could, with appropriate framing, stand up to the ecological concerns around the production and bottling of mineral and spring waters.

Let us return to the tasting room in Issy-Les-Molineaux and the lone bottle of tap water amid the crowd of bottled waters from around the world. What happened next, after the experts put down their glasses of water and the video cameras stopped recording their reactions? This work, Elle pointed out, aimed to valorize Nestlé's waters. The bulk of the results circulated to the public through Nestlé's region-specific marketing teams (discussed later). However, the results also traveled in another direction: via boardroom presentations to the highest-level stakeholders in the company. Gilles recalled that, after the completion and compilation of the results from the expert tasting, "we went around and talked to the different marketing teams and we had this presentation explaining the different categories . . . [we] did some blind tasting with them so they could wrap their heads around it."[21] In total, Gilles led three tastings: one in the Middle East, one in Paris, and one for the board of Nestlé. "That [last one] was kind of daunting," he recalled.[22] This approach to tasting did not remain in Issy-Les-Molineaux. Instead, it began to circulate through the company, a piece in a much larger game in which commercial water production collides with stock market economics.

To valorize bottled water in a world where tap water was making a surprising comeback, the team employed dramatized experiments to highlight the stakes of taste. Gilles, tasked with the process of teaching this new method of tasting to stakeholders, remembered explaining the process of categorization and co-tasting before then having the stakeholders themselves taste. "We tasted all the waters together and I tried to get them to talk about [the waters] and see if they could tell me which category each water belonged in, and also if they could identify which one was the tap water, because there was one in there that was tap water."[23] As with the little "trap" used at the initial tasting, Gilles told me that in every subsequent tasting, "every single time people are able to recognize tap water, even filtered tap water. That's the interesting thing," he mused, "you hear a lot, 'Oh tap water's the same thing,' but if you actually take the time and smell what's in your glass, you can pick it up right off the bat, and you know it's not the same thing."[24]

Tasting is a way of making molecular evidence perceptible. By offering a miniature version of the expert panel to stakeholders, complete with the "little trap" of a bottle of municipal water, Gilles used an individual's gustatory experience as the necessary proof of the value of Nestlé's waters. In essence, he created a theater of taste. In using the term *theater of taste*, I am riffing on Bruno Latour's concept of the theater of proof: the use of "dramatized experiments that allow spectators to see the phenomena [being] described in black and white."[25] The marked ability of individuals to recognize tap water, even when anonymized in a bottle, "proved" that bottled mineral waters were different.

The theater of taste validated marketing budgets and assuaged frantic shareholder concerns over the anti-bottle environmental messages promoted by municipal water producers throughout the Western world. In this politics of valorization, tap water, even when placed in a bottle, was still dramatically different from mineral water.[26]

## EDUCATING THE MASSES

Defining how to taste water did not stop at the boardroom. The communications team used insights from the tasting to develop print and web-based educational materials aimed at teaching consumers how to drink their water like a connoisseur. The educational materials instruct self-guided trainees via image and text to first check for the correct temperature (11–13°C/51.8–55.4°F for still waters and 8–10°C/46.4–50°F for sparkling). Next, trainees are instructed to fill a clean, odor-free, straight glass one-third full so that they may look for impurities before placing the glass under their noses. Imagery on the website further illustrates this procedure, showing the glass to be more like a whiskey glass or tumbler, shaped to encourage interaction between the olfactory cavity and volatiles without concern for temperature transfer from the body (e.g., there is no stem). Once the glass is under the nose, trainees are encouraged to "breathe deeply and regularly several times" with their eyes closed so that visual interruptions are minimized.[27] Tasting begins when a mouthful of water is placed in the mouth and allowed to "initially linger" before the water is "swilled around the mouth" and then held "over the back of the tongue before swallowing." Trainees are told that this will enable them to assess the water's "acidity, [taste], structure, lightness, and mouthfeel." The tasting does not end here. Readers are next instructed to evaluate both "balance and length of finish" by placing more water in the mouth. "Let it lie on the tongue," the instructions say, and then "breathe in a small amount of air through the mouth and breathe out through the nose."[28] Try it! Like any other self-directed trainee, you'll experience how dégustation means retraining your body away from its everyday habits.

Taste education is time intensive. It doesn't lend itself easily to mass communication projects. One might imagine that to align with the goal of making water tasting accessible to a wider audience, Nestlé's offered taste education needed to go beyond the distinction of connoisseurship. But it didn't. The terms the team generated reflect the exclusive side of the wine world characterized by elevated, reified, descriptive language.[29] Nestlé's tasters categorized the waters into types, producing "a new menu of six mineral water families . . . to suit every palate": Light & Pure, Rich & Revitalising and Bold & Distinctive for still waters; Soft and Elegant, Lively & Thirst-quenching, Vivacious & Energizing for sparkling.[30] None of these descriptors reference objects one can easily share. Instead, the descriptors reflect what literary critic Sean Shesgreen terms the "turn to poetic expression" in wine language embraced by those outside of academic wine research.[31] Learning to taste water within this framing transitions the step of sniffing one's water from an everyday practice to a classed activity. In the process, everyday expertise is discounted in favor of a form of expertise that would, ideally, push one away from a plebeian taste for tap water.

Unlike the taste and odor wheel generated for waterworkers through the Lyonnaise-AWWA Research Foundation collaboration discussed in chapter 3, Nestlé's public-facing tasting notes and instructions erased the possible presence of any off-flavors. Instead, the tasting notes present Nestlé's waters as carefully protected from human activity by layers of geological sediment. These individually distinct waters are "the product of [their] underground geological journey" and need to be "handled with care," protected and safeguarded for the future.[32] The tasting notes argue that bottled water is exactly as nature made it.

By adopting the language and practices of wine critics and connoisseurs into their taste education program, Nestlé sought to resituate both waters and consumers. Mineral water, through tasting, became a luxury product deserving treatment like the finest of wines. Archival research on France's mineral waters shows a striking similarity between perimeters of protection drawn for mineral water sources in the late 1800s and early 1900s and the Appellation d'Origine Contrôlée regulations that define specific geographic regions of France as having unique, legally defined terroir (e.g., champagne must come from the Champagne region).[33] The perimeters of protection drawn for mineral waters shielded spring owners from other mineral water entrepreneurs. Nestlé's argument for the terroir of water in its portfolio builds on the history of how bottled water transitioned from medicine to food; by situating its waters as having terroir, the company's marketing team aimed to protect Nestlé from ecological arguments against bottled waters. Taste education transformed consumers into apprentices learning to know their waters like they should know their wines. In so doing, Nestlé situated consumers as co-guardians of the lands producing water's terroir both for themselves and for future generations.[34]

Dégustation, and its accompanying language of valorization, was not the only type of tasting going on at Nestlé. Outside of the marketing offices and boardrooms, another type of tasting flourished: that of the scientist using FPA and the water tasting wheel to manage water quality. Like Lyonnaise, Nestlé has an entire research center in eastern France devoted to studying and managing the taste of water produced throughout its worldwide network. And like Lyonnaise, the ultimate goal is to identify off-flavors before they made it to consumers. "At Vittel," Elle noted, "the work is completely oriented towards the aspects of quality. There we have panels, we had pan—well, they still exist, but they are less active than in the past. There are expert panels that get together, I believe every week, to test the waters and vary the samples that come from all around the world and verify the quality."[35] Experts screen in bulk for the "different types of defaults that water can have with different odors and tastes," a methodological companion to the microbiological screenings the company carries out to ensure the water's safety.[36] One of the core reasons for the sensory screening, Elle pointed out, is to capture defects: "They do [screenings] with a very precise table and say 'Ah, well, there is an odor of this, and there, there is a taste of that' and this [odor or taste] shows for example, if there is a small taste of apple—that wasn't necessarily the case—but a taste like apple that could make them think that there was a problem with the type of plastic used, with the acetylaldehyde that was used [while making] the plastic."[37]

Scientists at Vittel also used taste and odor to alert producers to faults. Those faults might be in the water treatment process (critical for waters like Pure Life), in plastic bottles, or in the amount of time that a water had been stored. Vittel's scientists' use their senses to shield consumers from the industrial aspects of water bottling. As Elle put it, taste and smell said: "Stop production, you need to verify all these things before sending [the water] to the consumer."[38] From Elle's perspective, the entire focus of the Vittel research center was quality, but not the type of quality one could teach consumers via a website or transport to a boardroom as part of a theater of proof enacted via taste. Instead, the quality sought by the corporate scientists at Vittel, like that of the scientists at Lyonnaise or the MWD, was the quality of nothingness.

The tasting work done by the Nestlé team, and subsequent website and print information, paints a specific form of relationship between environment and corporation. The tasting situated Nestlé as a core protector of the environment, ignoring the role that Nestlé Waters (and one might say, all bottled water producers) plays in the extractive economies responsible for producing the lightweight plastic bottles that then circulate in the environment for millennia in increasingly small particles.[39] Teaching dégustation meant teaching consumers to prioritize terroir rather than pay attention to the entire political economy of bottled water production. In the process, the work of making a bottled mineral water that doesn't arrive on the store shelf with minerals that have precipitated out, or even of monitoring microbiological activity, disappears into the background. In only pointing toward the biogeographical and physical histories of water, Nestlé obscures what happens between the moment when a water is found underground and when it ends up in market. This is not to say that such modes of attention calling don't exist; they can be found in quality reports, for example. Rather, my aim is to point out that the stories that emerge through dégustation prioritize attention to long-standing understandings of the relationship between earth, food, and flavor at the expense of more recent environmental impacts of water exploitation. Attending to terroir makes it easy to miss that the systems and ways in which bottled water is produced are, like municipal water, deeply technoscientific.

## TASTING NATURE IN THE STREETS

Bottled water producers were not the only entities mobilizing taste to frame attention. In the early aughts, municipal water producers in large cities such as New York and Geneva began conducting their own theatrical demonstrations of taste in the streets and other public places. During the summer months of 2011, 2012, and 2013, Eau de Paris, the municipal water provider for the city of Paris, began running its own theaters of taste around the city as part of a new marketing campaign aimed at changing consumer conceptions about Paris's tap water.

In 2010 the city had regained control of water management and distribution from Lyonnaise des Eaux and the Compagnie Générale des Eaux. Taking back Paris's water, for the internal and external team tasked with the process, called for recapturing the hearts, minds, and mouths of citizens. Catching consumer attention started where water tends to spend a lot of time: underground. Bright pink posters with a simply painted grey-blue tap coming from the left of the image greeted Metro riders descending into the belly of the city. Instead of a faucet handle, a corkscrew offered access to the water. Viewers were invited to "Ouvrez un grand cru" (Open a grand cru) and "boire l'eau de Paris" (drink Parisian water). These posters, affixed alongside the normal litany of advertisements for furniture shops, concerts, and movies, stood out not only for their bold colors and childlike simplicity but also for the intrinsic shock value of the message: Paris's municipal water deserved to be savored and valued like the finest wines.

Grand Cru wines are grown on lands or by houses known for repeatedly producing wines that experts deem worthy of the highest prices. The classification, which dates back to 1855, indicates to buyers an exceptional quality that justifies premium pricing. This relationship between where a wine is produced and bottled, its taste, and its price has drawn sustained attention from researchers and wine experts alike as they try to understand what properties make a wine a "grand cru."[40] Simply stated, Grand Cru wines are the most valuable of French wines.

An internally conducted survey informed leaders at the newly public utility that approximately half of Parisian consumers did not trust the quality of their tap water. The "Ouvrez un grand cru" campaign, which ran through 2015, sought to directly assuage those fears.[41] Elisabeth Thieblemont, then director of internal and external communications and the one credited with the genesis of the campaign, put it this way: "You tell me you are not reassured [about the quality], I will show you the quality."[42]

The 2012 campaign called attention to the wide availability of "good to drink" water throughout the city. Already existing Wallace fountains were temporarily boxed up and then publicly "unboxed," revealing the "grand cru" within. Consumers were then invited to compare Paris's tap water with Evian and water from the minimally mineralized underground source of Albien.[43] This approach reflected a trend seen throughout Europe since the 2000s of municipal water providers confidently marketing their product in competition with bottled water.[44] These early public theaters of taste, like the theater of taste used in Nestlé's boardroom, often used what is known as a blind taste test.

Blind tasting, also referred to by sensory scientists as the "paired preference test," posits that tastings done without the accompanying informational bias of labels, bottles, and all the other cultural baggage that accompanies those external parts of a consumable good better allow basic truths about a substance to emerge.[45]

In contrast, in 2013 Eau de Paris did away with comparative, blind tastings. This resituated the tap water of Paris as a product to be evaluated on its own merits. The 2013 "Ouvrez un grand cru" campaign aimed to educate *all* consumers in Paris, be they longtime residents or newcomers, about the quality of Paris's water. Those I spoke with understood their task as especially focused on reaching out to two key groups who regularly consume bottled waters in France: the rich and immigrants. "The rich buy water in bottles without second thought, and the poor [immigrants] do not believe that the water is potable," one worker told me. Temporary booths set up in public markets, in parks, and at train stations helped achieve the goal of at least offering a drink to the majority of Parisians.[46]

Outreach teams included a mixture of Eau de Paris employees from the communications team or scientific team. A group of young, enthusiastic, outsourced workers trained on Eau de Paris's core educational aims rounded out the team. As pedestrians passed by, the team offered a drink of water. The invitation to have a drink of water set the stage for longer, informal conversations. Follow-up conversations included asking what drinkers thought of the water; providing educational commentary and materials in the form of pamphlets, magnets, and flasks; and answering any questions.[47] Education about where Paris's water came from and how it was cared for played a central role: on my first day of observing, a technical volunteer explained to me the different sources of water feeding Paris. But instead of just telling me, he opened the *Ouvrez un grand cru* brochure, flipped to pages 4 and 5, and used the simplified map to illustrate that surface and underground waters largely came from locations distant from Paris, water "unsullied" by urban life. He then showed me how the waters come from a variety of surface and underground sources that are mixed together and then redistributed to a limited number of quarters: The water in the 16th arrondissement comes from river and groundwater, while the water consumed in the 17th, an arrondissement just to the north, is from groundwater alone.

Despite the different provenances of Paris's waters, all of the water supplied throughout the city has been standardized to the same average mineral composition. The mineral composition of Paris's water made up a core part of the 2011–13 outreach efforts. Eau de Paris labeled public fountains throughout the city with 2-by-3-inch plaques that highlight the mineral content. They included the numbers in marketing materials. The mineral content label even appeared on the side of arrondissement-specific carafes sold throughout the city. The outreach brochure similarly pointed out that consumers could expect the water in their mouths to taste of 90 mg/L calcium, 6 mg/L magnesium, 20 mg/L sodium, 2 mg/L potassium, 220 mg/L bicarbonates, 30 mg/L sulfates, 20 mg/L chloride, 29 mg/L nitrates, and 0.17 mg/L of fluoride. Parisian drinkers who were already adept at evaluating the potential flavor of a bottled mineral water based on such numbers would immediately recognize that Paris's water was lightly mineralized, suitable for everyday drinking.

Such widespread use of numbers to inform potential consumers about Paris's water quality is striking; European Union regulations did not require Eau de Paris to label public fountains with the water's mineral content. It took time, money, and effort to label the fountains around the city. What were these numbers doing? Despite being singular, numbers estimate a particular form of truth and in the process help define what deserves attention.[48] On the most basic level, the numbers partially demonstrate that Eau de Paris produces water that falls within European Union regulatory standards, although few people would readily have such numbers at hand.[49]

The numbers also communicate palatability. The mineral water producers of nineteenth-century France intimately knew this, proudly displaying the mineral content of their waters not only on bottles in accordance with legislation, but also in marketing materials. The measured iron content, for example, of Bussang let consumers know not only that this mineral water would work to restore the blush to a daughter's face, but also that the water would *taste* like it worked. Mineral waters are still labeled with their mineral content, although the mineral water producers of the twenty-first century have shifted the conversation from specific ailments toward broader issues like digestive health (a water with high magnesium content would be appropriate) or environmental concerns about contamination.

In contrast, until recently many municipal water producers have largely ignored the effects of mineral content on taste and consumer health, focusing instead on the ongoing struggle to maintain safety throughout large distribution systems.[50] By voluntarily highlighting the mineral content of Paris's tap water, Eau de Paris linked its current work with nineteenth- and early twentieth-century efforts by physician-chemists to determine mineral water quality. As George Weisz compellingly argues, French physician-chemists understood that quantification of mineral content could help them link their empirical observation of a water's clinical impact and scientific fact.[51] By labeling the mineral content of its water, Eau de Paris consciously invited consumers to attend to the city's tap water within the larger historical discourse surrounding mineral waters: How will this water contribute to your health? Is this a water that should be valued, legally protected by both the rule book (*cahier de charge*) and a perimeter of protection, in the same fashion as France's medicinal waters or finest of wines?[52]

By adopting and applying the mineral-content labeling system French consumers were already familiar with to Paris's municipal water, Eau de Paris's communications team repositioned the utility's product as equal to—or even better than—mineral waters. In fact, the mineral content of Paris's tap water closely reflects the mineral content of Evian, the water most commonly used in France as the "standard" for neutral drinking water. Although on the surface the relabeling of Paris's water with numbers appears to be another marketing tool, it can also be seen as an inflection point in the making, a slow reworking of the Parisian taste for water.

There are other types of numbers associated with municipal (and mineral) water production that do not appear on the posted labels. Bacterial count reports require a visit to a municipality's website. Finding such reports for bottled waters proves a more complicated sleuthing endeavor. Similarly, levels of pharmaceuticals or pesticides are not communicated on labels; in fact, the *Ouvrez un grand cru* booklet handed out in 2013 places them at zero.[53] The company thus unfortunately missed a key opportunity for communicating with consumers about the need to protect the watershed.[54] The charisma of the visible numbers as used in the "Ouvrez un grand cru" campaign, like the training offered by Nestlé, focused consumer attention on taste. Eau de Paris's decision to mobilize a theater of taste supported by numbers makes sense within the larger contest between bottled and municipal waters—a contest largely carried out in the realm of taste and Taste. However, had Eau de Paris also highlighted the less-visible numbers, such as those found in raw water, it could have more easily surfaced how industrial, agricultural, and urban activities had impacted and were negatively impacting the quality of water entering purification plants. Primarily focusing on taste directs attention to the finished product, not its origins.

## VALORIZING WATER

"Water is an alimentary product, good to think with, and good to drink," Elisabeth Thieblemont, the director of Eau de Paris's communication team, remarked.[55] Quality, for Thieblemont, manifested through tasting. Despite the fact that a minute fraction of the municipal water used in cities is actually ingested as a food, Thieblemont recognized that highlighting the taste of Paris's water offered a way for the utility to quickly and effectively link perception, ideology, and behavior.

The concept of valorization—the action of "fixing the value or price of some commercial commodity" or of "making valid" (*mettre en valeur*)—appeared repeatedly in printed materials and in my interviews and casual conversations.[56] Valorization in the late twentieth and early twenty-first centuries is proving especially critical as water infrastructures age and utilities begin passing on replacement costs to inhabitants, both as consumers and as taxpayers. Eau de Paris's aim to valorize its work, and water, thus reflects a contemporary view of water as an undervalued material located within an undervalued system. For Thieblemont, drawing from the language of business made perfect sense—the utility did not have the funds to go head to head with bottled water companies, but it certainly could draw from their techniques in valorizing Paris's municipal water. "When I started, Eau de Paris wasn't doing a lot of communication with its customers," she reported.[57] "My goal was to reassure the population about Eau de Paris, both as a public service, and as a provider of quality water," she continued. To do this, she proposed the utility adopt a brand strategy: "That is why we changed the logo. It is now 'Eau de Paris. A public service.' Not 'Eau de Paris, a public service.' The period is important." In distinguishing between the function of the company, the name of the product (the water of Paris), and the management company (Water of Paris), the communications team explicitly linked to the larger political discourse of the then-ruling Green Party: public goods are better served by public hands, not by outsourced private entities.

The "Ouvrez un grand cru" campaign sought to provide transparency about the production techniques of Eau de Paris by bringing its scientists and waterworkers to the people (along with the core team of event people). It also sought to make Eau de Paris a recognized, trusted brand. Water, as *the* product produced by Eau de Paris, was key to brandmaking. Unfortunately, from Thieblemont's viewpoint, the industrial production techniques used to "make" municipal water, hidden behind factory doors or closed gates, distanced consumers from producers: consumers had no idea who produced their water or how they produced the water. "We wanted to create proximity where there is thirst," she told me. Creating proximity meant offering free educational tours of facilities. Creating proximity meant opening new, more visible, public fountains—both still and sparkling—around the city. Creating proximity meant doing definition work about taste in the most public of spaces in a city: the street.

Successfully training someone to pay attention through taste relies on definition work: the work of setting boundaries of understanding. Definitions signal what problem(s) are worth examination and identify what receives attention.[58] Defining shapes how individuals *and* institutions notice things, as well as what they notice; definitions create infrastructures for organizing knowledge.[59] Doing definitional work on the street in 2013 relied on a simple formula. Volunteers and workers invited those passing by to try the water through the simple question: "Would you like a glass of water?" or "A glass of water?" The water's provenance remained left out of the initial invitation, although if someone asked where it came from before trying the water, they received an answer. Almost everyone who stopped tried the water, pausing momentarily to drink before moving on. When crowds were small, a team member would ask: "How is it?" Over and over people responded, "It's good!" The volunteer would then respond with, "It's Paris's tap water" (*C'est l'eau de Paris/c'est l'eau du robinet*).

In adopting the language of connoisseurship while teaching consumers how to "transcend the agreeable to make taste judgments about what is good and suitable," the "Ouvrez un grand cru" campaign created a contradictory set of tensions.[60] The familiar, albeit at times inaccessible, universe of oenology was linked to deep-seated cultural beliefs about terroir. By situating water like wine, the campaign deliberately highlighted that water, too, was made through careful labor to produce a high-quality product. The invitation to consumers to consider their municipal water alongside the finest of French wines valorized Eau de Paris's workers and the work of the newly "liberated" public utility. Yet as Bénédicte Welté, a senior scientist for Eau de Paris, pointed out, the language of connoisseurship also reified water. "Water is a public good," she dryly remarked, "not something that is inaccessible to everyday people." For her, speaking of water like wine potentially alienated a large portion of Parisian consumers and undermined the ability of the utility to carry out critical conversations about water accessibility.

Unlike earlier outreach campaigns, which offered participants a blind tasting between tap water and bottled water(s), and at times mineral water, the 2013 "Ouvrez un grand cru" campaign simply offered tap water. The decision to transition away from a paired preference form of blind tasting matters: the tasting retains a theatrical, charismatic form to viscerally argue for the value of municipal water. Simply sharing a single sample enabled a different conversation to ensue about how to "care" for or attend to tap water in the home in ways that would allow it to taste as good as in the streets. In so doing, it invited consumers into the process of making their tap water's terroir.

The outreach campaign's name implicitly invited participants to not just drink but also taste with the goal of building up positive environmental behaviors linked to sensory acuity—the ability to go beyond liking or disliking in favor of other values.[61] Like Nestlé's consumer-centered web page and publications that offered self-directed taste education, the Eau de Paris tastings ideally positioned consumers as careful tasters who would pause, observe the water's clarity, smell the water, and carefully consume, all the while judging the quality of the water in their glasses. Unlike Nestlé, Eau de Paris did not seek to educate consumers on how to properly "drink" water. Instead, consumers were left to perform consumption as they wished, a leveling of the social act of dégustation that nonetheless enrolled bodies into the sensory labor of determining the value of Paris's municipal water.

Consumer's performances varied widely. Some explicitly tasted their water like one would taste a wine. For example, one man, when asked by the video team how he found the water, sipped, swished, and then responded, "This is a refreshing water . . . with a taste, well it is not completely neutral, which is always notable in a water to not have a neutral pH, one can feel/smell the same taste as a mineral water."[62] He specifically performed connoisseurship in his actions, trying to adopt the language of wine tasting into this new situation. Many found it harder to find words beyond "good" to describe the water yet still followed the pattern of looking, smelling, and tasting. Many, especially on the hot, sunny Sunday in the Parc Buttes-Chaumont near the end of the campaign, simply downed the proffered glass and moved on.

Not all of the responses were so positive. One elderly woman informed the volunteer team and surrounding onlookers that her dog refused to drink anything other than Evian, and she followed suit, implying that her pet could sense danger that she could not. Another, when interviewed by the video team, cautiously responded, "This water is good. At the taste level, but pollution is transparent, even so I would not like—but at the level of taste, it is perfect."[63] Both of these women did not trust the processes required to produce municipal water, distrusting either the base product or the technological interventions that render it potable.

Whether their responses were positive or negative, a significant number of drinkers replied to learning that the water they had tasted was Paris's tap water with a variation on, "Excuse me, but we do not have *this* water at our house." Across the locations I observed and volunteered at, consumers repeatedly responded with surprise that the water being served did not taste at all like the water coming from their taps. The discrepancy between the taste of water from the tap at home and the water served in a white paper cup at the market, train station, riverfront, or pride parade was the crucial balancing point of Eau de Paris's outreach campaign.

Team members had a specific strategy to further invite people into becoming partners with the utility. Although the Eau de Paris team certainly sought to raise awareness of the impact building pipes have on flavor, the main verbal focus of the conversations I observed and participated in was how an individual could manipulate and manage the taste of the tap water in their home; As such, the Eau de Paris team offered guidelines for managing common off-flavors found in homes. Sometimes this included highlighting the impact pipes in buildings can have on tastes and odors. More often, the invitation to consumers to take part in redefining the taste of tap water in the home occurred through a focus on managing the perceptible reminder of water's potential dangers. One single category of off-flavors consistently appeared in conversations: the odor of eau de javel (chlorine) was missing from the water served during the outreach campaign. "But where is the chlorine?" many asked during the days I volunteered, skeptical that the water being served was actually the water in their pipes. Chlorine's imperceptibility at the tastings, a result of the water being served out of uncovered carafes (the water was directly taken from the municipal system), troubled preexisting conceptions of the quality of Paris's tap water.

French consumers are notoriously sensitive to the presence of chlorine in their water, simultaneously seeing it as a necessity for safety and a sign of water not fit for human consumption.[64] Bénédicte Welté, told me that for Parisians, the multiple uses of tap water complicate Parisian's views of the quality of the water. "When we ask them 'Do you find the water good or not?' they say to us, 'No. The water is not good, because I have deposits in my pans.' And often, because the water tastes of chlorine. Those are the two responses we get."[65] To respond to these concerns, Eau de Paris combined the theater of taste with education about how consumers could act on their own future sensing bodies to make the tap water they drink at home taste as good as it did in the street.

One only needed to make a few small behavioral changes to get tap water to taste good at home. Eau de Paris team members suggested letting the water run for a minute in the morning to wash away any stagnant water, allowing the water to breathe for a bit so any residual chlorine could evaporate, cooling the water to increase its freshness, or simply adding a bit of lemon. Any one of these techniques, used singly or altogether, promised to give water as good as that served on the streets by the Eau de Paris team in a white paper cup. These small practices, taken together as a whole and distributed across the metropolitan area, offered drinkers the ability to join in a bottom-up form of sensory labor in which they collaboratively valorized the decision to bring Paris's water back into the public sphere. By showing drinkers how to appropriately pre-care for their water before consumption, the Eau de Paris team suggested that Parisians could remove the remaining perceptible markers of the industrial techniques used to safely deliver water across thousands of miles of pipe. As people encountered the good taste of Paris's municipal water and made their observations public, they opened themselves up to the possibility of becoming partners with Eau de Paris in defining and valorizing Paris's water.

## THE CONSEQUENCES OF TASTING WATER'S TERROIR

The process of training someone to taste changes how people use their mouths and noses. When juxtaposed, the theaters of taste conducted by Eau de Paris in the street and Nestlé Water's marketing team in the boardroom demonstrate a shared assumption: that as people learn to pay attention to sensory cues, they gain access to previously hidden realities and actionable "truths" about perceptible materials circulating in the environment. By framing their waters as having terroir, Nestlé and Eau de Paris understood people's experiences with the theater of taste as capable of shifting attitudes and decision-making in the competition between bottled and tap waters.

The performance of tasting played a key role in the theater of taste. Nestlé invited its board members and anyone interested enough to drill through its website to move toward tasting like a connoisseur. In contrast, Eau de Paris invited participants to expand their everyday expertise to include caring for one's tap water at home. The work of teaching people to look, smell, and then taste mineral water as Nestlé's team aimed to, or of teaching people to let their water breathe before serving, or even chill it, is an inflection point. In this new realm, the work of smelling and tasting invites active, conscious sensory labor to replace the previously unconscious work of evaluating water. In doing so, it also resituates taste on the hierarchy of public-facing values.

I love the work municipal water producers have done to mobilize taste as a method for inviting consumers to reconsider how they evaluate quality.[66] Yet placing such theaters of taste alongside those of a bottled water producer invites one to pause and ask, "Why taste?" After all, regulatory authorities situate taste as a secondary standard. Taste and smell lack authority within the systems that have come to define quality through microbial counts and maximum contaminant levels. Focusing on taste resituates both the work of defending water's quality and of valorizing water in the mouths and noses of everyday consumers in a way that is constantly—and consistently—under threat due to water's many other uses.

The taste of water, when tied to terroir, carries the potential to call for better environmental management practices. But in practice, focusing on the taste of water simply situates water as always and only present and produced for humans or their preferred furry companions. Taste doesn't call for changing the upstream or downstream infrastructures that dump human-made contaminants into water. Instead, it invites a sly sidestep toward individualism, where what *my* mouth prefers ultimately is situated as what guides my relationships with others. And that move, unfortunately, obfuscates the larger question we ought to be asking: What environments do conscious forms of sensory labor, when done by everyday drinkers spread throughout a system, actually produce?

# 5

# ERASING PLACE

## Industrial Terroir in the Twenty-First Century

IN NOVEMBER 2019 the Scottsdale Arts Council and Scottsdale Water hosted their first-ever One Water Brewing Showcase, a brewing competition that invited brewers to use highly purified wastewater (also known as recycled water) to make a limited-edition beer. As part of Scottsdale Arts' Canal Convergence festival, an arts festival that annually brings thousands of inhabitants and tourists into the heart of the city during a "shoulder" season, the One Water Brewing Showcase offered ticket holders the chance to try unique beers from eleven local breweries. With each sip, drinkers found themselves tasting a speculative future not yet legally possible in Arizona—a future in which wastewater was immediately cleaned and returned to the municipal water supply system. Participating brewers acted as "water heroes," helping in the effort to "normalize attitudes towards purified recycled water."[1]

Arizona's need for "water heroes" emerged in the twenty-first century context of "water scarcity, drought and lack of access to clean water."[2] Increasingly, questions of raw water quality intersect with and shape questions of raw water quantity. Such questions of quality and quantity are especially pressing for arid regions. Locations like the desert Southwest of the United States, regions of northern India, countries around the Mediterranean, the Middle East, and Australia are all characterized by the World Resources Institute as having extremely high or high water stress.[3] Access to an adequate water supply of good quality shapes the current and future ability of arid and semiarid regions to prosper.

This question of water quantity and quality certainly hits close to home for me. Arizona's Maricopa County, my current home region, is one of the fastest growing metropolitan regions in the United States. Yet it has an average rainfall of only 8 inches (20 cm) per year. This means that meeting the water needs of the region's inhabitants (human and otherwise) is a pressing environmental challenge for policy makers and water providers. Recent record-breaking heat trends, multiple dry monsoon seasons, significant in-migration to the area, and ingrained habits of treating desert places like wetter climates mean that regulators, policy makers, and others responsible for supplying water to water-stressed regions like Arizona are constantly on the lookout for "new" sources of water to augment current supplies. Indeed, as I write this, the US Bureau of Reclamation has moved the lower Colorado River basin states of Arizona and Nevada, as well as Mexico, into a Tier 2 shortage level, activating additional cuts to the region's water supply.[4] For policy makers and water providers, dwindling water supply is not the only concern. Increasingly they find themselves addressing questions about "contaminants of emerging concern" not yet covered by drinking water regulations.[5] Contaminants of emerging concern include pharmaceuticals, pesky "forever" chemicals like the PFAS (per- and polyfluoroalkyl substances) found in nonstick coatings, and ingredients found in personal care products.[6] They are the chemicals that make their way into water but are not yet tested for, nor are they legally regulated in many places. The EPA recently issued a new list of contaminant candidates that includes PFAS, and the EU is currently deliberating restricting PFAS.[7] Compromised water quantity and quality have converged to spark a new set of possible water futures.

Water recycling, also known as potable reuse, sits at the heart of these possible water futures.[8] In this final chapter, I draw a line from the flavorful pasts explored in previous chapters to the present, where proponents of water recycling are taking industrial terroir to its logical end: an approach that will not just remove the perceptible traces of environmental variability, but will do so in a way that offers to unhook deeply held social and cultural associations linked to a water source's origin. Water recycling, which I've been collaboratively studying with Marisa Manheim since 2019, offers providers a pathway for transforming water, no matter its source, into whatever type of water providers need it to be. Despite the technological prowess at the heart of potable reuse, the technology remains in a liminal state, in large part due to public resistance to the idea of using wastewater as a source for drinking water. By examining this technology as it hovers at an inflection point, this chapter offers insights into how demonstrating quality with not just numbers, but also bodies through public tastings of blank-slated water—water lacking any perceptible markers of the natural environment—enacts futures. The future can often feel out of reach. Yet it is also a potential set of realities in the making. Within that framing, I am curious how the management of taste and smell in the present intersects with anticipated water futures and the sensing bodies imagined to populate those futures.

## POTABLE REUSE

Like many people around the globe, I stayed home over summer 2020. As temperatures approached 120°F (48.9°C), I and many other Arizona inhabitants began looking forward to the intense late-afternoon summer thunder and rainstorms, locally known as monsoons, that break the oppressive heat. Unfortunately, Arizona experienced a "non-soon" season in 2020. Non-soons are monsoon seasons with little to no rain. Wildfires raged across the state, wiping out massive numbers of native (and non-native) plants. The Arizona Game and Fish Department delivered water by helicopter or truck to rural catchments to help struggling wildlife populations. These increasingly normal events bring home the reality that the largely desert region is in a state of drought. According to experts, it has been for over twenty-four years.[9] Standing next to the empty banks of the Rio Salado makes Arizona's water future look grim.

There's a long history to how Arizona became as water stressed as it is. It includes attempts by settlers to turn the traditional homelands of the Akimel O'Odham and Piipaash peoples in the area now known as the Phoenix region into a Mediterranean landscape; agricultural and industrial extravagances; population growth post–World War II with the promotion of the Sunbelt region; and more recently, climate change.[10] By the 1940s, many parts of Arizona were operating on a water deficit, withdrawing more groundwater each year than rainfall replenished. As use increased, the amount of water withdrawn from groundwater sources greatly exceeded the amount of water being returned to the ground. Over the past eighty years, some areas of the state have seen significant drops in the surface level of the land (a phenomenon known as subsidence), as well as the abrupt emergence of deep cracks in the landscape.[11]

The state has worked over the last forty plus years to address these issues of water quantity, starting with passage in 1980 by the Arizona legislature of the state's Groundwater Management Act. This act required municipalities to maintain "safe yields" of groundwater by matching how much water they withdraw with how much water they return to aquifers; one can think of safe yields as being similar to maintaining a steady amount of money in a bank account over time.[12]

Water management specialists have put forward a range of practical and fantastic strategies to replenish aquifers. Examples include water augmentation (e.g., desalinization of brackish groundwater, reducing the number of trees in forests, groundwater recharge), improving current water use (e.g., conservation, reusing wastewater), and efforts to innovate in water augmentation (e.g., cloud seeding).[13] While all of these approaches rely on technological intervention, water reuse is the one that draws people's intimate experiences of smelling, tasting, and consuming into close interaction with water futures.

Water reuse is pretty much what it sounds like: taking water that's been used once and using it again. Many who live in water scarce regions practice informal forms of reuse such as using water from rinsing vegetables to water house plants. Indeed, using water containing discards is not new. Historically, what most water professionals and everyday folks now characterize as "wastewater" was pretty valuable stuff. It provided needed fertilizer for fields. It was the source of urea for tanning and for basic weapon making.[14] The advent of modern chemistry and the rise of germ theory undid much of that value.[15] Yet as issues such as water scarcity, depleted aquifers, and even conflict over who gets access to water increase, reuse of wastewater has gained acceptance as a valid "new" source of water. A range of municipalities now embrace reuse for agricultural irrigation, landscape applications, and power plant cooling.

Reuse falls into two categories: indirect and direct. In indirect reuse, wastewater treatment facilities return treated water to aquifers or other natural buffers. The water must remain in the aquifer or natural buffer for a defined period of time before it can legally reenter the water treatment and delivery system. In some cases, indirect reuse gets categorized as "de facto" wastewater reuse. This happens when treated wastewater is returned to a surface water source and then becomes an unplanned part of the raw water supplying a downstream water facility.[16] Indirect reuse has proved a politically palatable approach for introducing reuse to communities. For example, due to a Scottsdale mandate that golf courses use reclaimed water in 1989, golf courses partnered with the city to fund building of an advanced wastewater treatment plant; all courses in Scottsdale are now watered with reclaimed water.[17] Scottsdale then uses any excess water not used by the golf courses to recharge its aquifers, a move that the city points to as helping it reach "safe yield" levels nearly twenty years before mandated to by Arizona law.[18] In contrast, DPR sends the treated wastewater directly back into the water delivery system, either by putting the treated wastewater into raw water headed for a drinking water treatment plant or by blending the treated wastewater with finished water ready for distribution.[19] For technologically minded folks who think of water in terms of its molecules, it's not a big leap to go from purifying wastewater to levels where it can be used to safely water golf courses or recharge aquifers to purifying water to the point that it is considered safe for human consumption. The transformation of wastewater back into drinking water relies on a combination of different advanced treatment techniques grounded in the philosophy of, as Scottsdale Water representatives frame it, "treat, treat, and treat again."[20]

A series of technological treatments and analytical scientific protocols enable water workers to transform wastewater into drinking water. The DPR water produced by Scottsdale in its demonstration facilities is "an RO permeate that has been further treated with ultraviolet photolysis."[21] For those of us rusty on our chemistry or engineering terms, what that means is that the city's waterworkers take wastewater that has been treated to remove biological materials and nutrients, filter it, and then further purify it by passing it multiple times through membranes with really tiny holes. Finally, they treat the water with ultraviolet (UV) light. The UV light acts as a final guarantee of safety by breaking apart the bonds holding together any remaining molecules that aren't $H_2O$ and its derivatives.

However, the trip from wastewater to drinking water does not end with treatment by RO and UV photolysis. Technological treatments and scientific analyses are incapable of completely transforming wastewater into drinking water on their own. Laws, regulations, and permits govern the ability for a wastewater to be reclassified as drinking water in the absence of an environmental buffer. Brew challenges such as the One Water Brewing Showcase (2019; 2022, Arizona), and its predecessor, the Pure Water Brew Challenge (2015, Oregon; 2017, Arizona), only happened because Arizona's and Oregon's state Departments of Environmental Quality (DEQs) issued temporary or provisional permits loosening in-place regulatory restrictions.[22] These restrictions prohibited use of recycled water for direct human consumption without specific authorization.

Permits outline the scientific and technological parameters that define the final step in transforming wastewater into drinking water. For example, before DPR water can legally be served to any visitor to Scottsdale Water or be delivered to a beer brewer, it must successfully navigate the analytical tests required by the provisional permit enacted by Arizona's DEQ. At least a week before any tasting, utility employees must test the water daily for *E. coli* and total coliform counts (coliforms are a type of bacteria found in the intestines); if either test comes back with a "detected" result, then the tasting or delivery to beverage producer must be canceled or postponed. Utility employees must additionally test for twelve indicator parameters at least once within a seven-day period before serving the water.[23] A utility that fails to follow permit guidelines faces state and federal sanction.

Permits limit how DPR water can circulate. Permit drafts, draft fact sheets, and modifications related to the permit issued for the 2017 Arizona Pure Water Brew Challenge highlight regulators' hesitancy around DPR, as well as the liminal nature of the innovation. For the 2017 competition it was especially unclear what brewers would be allowed to do with the beer. Could they sell it to a bar? A restaurant?

The initial permit only allowed beer produced using DPR water to be "made available for tastings or sale directly by the breweries" or to be returned to the permittee, the Pima County Regional Wastewater Reclamation Department.[24] In fact, a modification had to be submitted to allow the brewers to sell beer produced for the challenge to retail customers; it was approved two months after the original permit.[25]

Scottsdale's efforts to build the 2019 brew festival included expanding where DPR could be served, albeit on a limited scale. At the time of writing, the provisional permit issued to Scottsdale Water by the Arizona DEQ divided usage of the water coming out of the DPR system into two categories. The permit's first category limits the utility to serving DPR water to fifteen hundred people per year, including drinking samples consumed by staff. The second category allows DPR water to be widely distributed to off-site beverage companies to produce "water-based beverages" (e.g., beer).[26] By limiting the amount of water that can be served "in-house" the permit circumscribes Scottsdale's ability to demonstrate the quality of DPR water to large groups of people.[27] In contrast, the permit provides no limit beyond the system's capacity for DPR produced for off-site beverage companies to further process it into "water-based beverages." DPR's uneven circulation—readily available to those over eighteen who drink alcohol at rare events; extremely limited annual availability for the rest of the population—shapes who gets the chance to encounter DPR firsthand during the liminal phase in which cities and legislatures have not yet decided to move forward in adopting the technology.[28]

Technological, scientific, and regulatory infrastructures combine to ensure that DPR water is safe, or at least safe as determined by the current state of scientific knowledge, technological capacity, and regulatory structures. Each establishes a different set of hurdles designed to demonstrate that producers do not need the natural environment to serve as a cleansing or diluting intermediary between treatment and humans. Instead, analytical scientific practices and technological know-how stand in for the work of nature. State and federal regulatory infrastructures act as an additional buffer. Water produced through the DPR process is more stringently treated and monitored than any other publicly available drinking water source, bottled or municipal. As Scottsdale Water's current director, Brian Biesemeyer, recently pointed out, there is "no logical reason not to consume [DPR] water."[29]

## EMBODIED REACTIONS AT THE INFLECTION POINT

Despite the fact that there may be no logical reason to not consume DPR water, efforts to implement the process, largely located in urban centers in water-poor regions (at least in the United States), have a history of stumbling.[30] Ask someone if they want to drink water recycled from the toilet, and they most likely will say no (unless you tell them it's happening on the space station and they get to be an astronaut; that certainly seems to shift the equation!). This refusal, many researchers say, is due to the "yuck" factor.[31]

What is the yuck factor? Anthropologists and theorists such as Mary Douglas and Julia Kristeva have long been curious about what makes something yucky, disgusting, or distasteful.[32] They point out that moments when people think "yuck!" are often tied to socially and culturally defined risks. Sometimes the risk assessment activated by yuck is physiological, such as when someone recoils from a bitter substance. Sometimes the risk assessment activated by yuck is psychological, such as when a caregiver physically recoils from an object a child picked up out of a garbage can. As demonstrated by the range of fermented food lovers and active communities of dumpster divers, yuck is not just innate, it is also learned.

People often report opting out of using tap water as their drinking water because they find the water yucky. Scholars interested in the turn to bottled or packaged water in countries with robust municipal provisioning systems routinely point to personal taste preferences as central to the bottled water industry's meteoric success at the end of the twentieth century.[33] Almost a hundred years earlier, Harvard's George C. Whipple pointed to a similar challenge: "It is a well-known fact," he noted, "that in cities which are supplied with water which is not attractive for drinking purposes, large quantities of spring water and distilled water are sold," and purchases of personal filtration systems increase.[34] For contemporary water providers, the voiced and silent opting-out of inhabitants in their districts can often appear irrational, prioritizing an aesthetic reaction to the tastes, smells, textures, or temperature of their water over trust in authorities, monitoring agencies, or science. I suggest we consider this opting-out of using municipal water as drinking water within a larger framework of hesitancy, refusal, and suspicion, rather than simply approaching it as an irrational aesthetic quirk to be educated away.

Using ideas of hesitancy, refusal, and suspicion requires a brief detour away from water into the world of vaccinations. Since approximately 2011 the medical community, social scientists, and policy makers have used the term *hesitancy* to describe where people fall on a spectrum of acceptance or rejection of vaccines.[35] In her ethnography of how people in Barbados accepted or rejected the human papillomavirus (HPV) vaccine, Nicole Charles points out that conversations about vaccine hesitancy in the medical field often "discursively misconstrue hesitancy as just a delay or refusal purportedly rooted in ignorance." Charles argues that calling the forms of opting-out she observed hesitancy or refusal overlooks the role individual and community memory, as well as emotional response, play in shaping Barbadian's responses to the HPV vaccine. Instead, Charles proposes "thinking with suspicion" as a better fit for understanding why people in Barbados exhibited resistance to the HPV vaccine. Suspicion, Charles notes, goes beyond hesitancy: it highlights the role of embodied, affective experiences like "gut feelings, emotions, and colonial residues of trauma" in shaping how people in Barbados responded to campaigns promoting HPV.[36]

Suspicion, as Charles is careful to note, is a concept specific to Barbados; it is embedded in the experiences Barbadians have had with "colonial modes of scientific knowledge" in action in the English-speaking Caribbean.[37] Despite this specificity, I think suspicion's attention to embodied, affective experiences, and how those experiences intersect with community memories of institutional harms or failures, offers a useful framework: even if suspicion cannot be directly used as a mode of analysis, thinking alongside the concept calls attention to the work being done through present-day embodied experiences in trying to shape how tap water's *anticipated future* aesthetic characteristics will interact with individual and community memory. As such, thinking alongside suspicion provides a road map for considering how efforts to convince consumers and policy makers that DPR is a viable and desirable innovation might play out in future sensing bodies.

## MAKING NEW MEMORIES: TRANSFORMING DPR FROM A LIMINAL TO AN ESTABLISHED TECHNOLOGY

DPR has existed as a known approach to obtaining "new" water sources since the late 1970s. Despite this, it may better be understood as a "liminal innovation." Liminal innovations, as Mayli Mertens argues, are "defined by continuous transition on both sides of the threshold between experiment and implementation"; they are "not yet used, yet 'not used.'"[38] The number of active sites doing DPR in the United States demonstrates the liminal state between "not yet used, yet 'not used'": the first US-based experimental DPR plant opened in Denver, Colorado, in 1979. It operated for fourteen years. The subsequent years witnessed a spectacularly failed attempt in San Diego, California, to implement reuse.[39] As of this writing, two cities in the United States, Big Spring, Texas (2013), and Wichita Falls, Texas (2014), have actually implemented DPR in their portfolio. Wichita Falls produced DPR water until 2016, when it returned to indirect reuse.[40] Despite the small number of currently active DPR systems, a significant number of cities are actively exploring a future with the technology. They are doing so through a variety of approaches, including development of master plans, early efforts to seek stakeholder input, public-facing education and outreach, and demonstration treatment systems ("treatment trains"). DPR is used, but not at scale. It is a technology teetering at the inflection point of widespread adoption.

DPR's liminal nature means that the technology's proponents are still working to figure out how to help the technology diffuse into legal and social realms. As a reminder, scholars interested in the adoption of innovations see diffusion as "a kind of social change."[41] Some, such as the manufacturers of RO filtration membranes, hope for a literal diffusion of the technology into regular usage in municipal water treatment facilities around the globe. Others dream of a massive public diffusion of the idea that technological, scientific, and regulatory infrastructures are capable of standing in for the environment.[42] Community hesitancy, resistance, or something akin to suspicion threatens proponents' ability to make such dreams reality. By partnering with beer brewers to produce tasty beverages, proponents of DPR, like the food scientists discussed in chapter 3, aim to create new, positive memories in the future sensing bodies of consumers—consumers who soon may be asked to support infrastructural or legislative retooling of water provisioning. Organizers of tastings seek to sever the affective connections between past experiences and expectations around wastewater so that DPR can finally transition out of its liminal state.[43]

For people attentive to numbers, the choice to use beer and its brewers as ambassadors for DPR seems odd; it can take between eight and twenty-four gallons of water to produce one pint of finished beer.[44] On the other hand, beer is 90–95 percent water. In fact, beer quality and style depends in part on water quality. A water's mineral content, pH, and hardness have historically shaped the differing regional flavors and characteristics of beers.[45] While in the past brewers relied on ground and surface waters located near their brewery, contemporary beer brewers in the United States largely draw on municipal water. Brewers trying to produce a consistent product are most successful when they know what is in the water they are brewing with and can adjust for variations. For brewers interested in brewing different *types* of beer from the same municipal water source, and especially for brewers from regions where water sources vary seasonally, installing a water purification system such as an RO significantly improves their ability to make any style of beer, and to make it consistently.[46] "Bad" water makes "bad" beer.

If "bad" water makes "bad" beer, then the inverse—that good water makes good beer—explains the choice of beer as a launching point for making an ingestible argument to drinkers about the quality of DPR water. Organized efforts in the United States to use beer made with recycled water first started in 2014. That year Clean Water Services, a wastewater treatment organization that primarily serves Washington County, Oregon, partnered with the Oregon Brew Crew homebrew club to invite home brewers to make beers out of recycled water. The initial water offered to brewers was 30 percent effluent—essentially, Clean Water Services obtained water from the river downstream of its discharge point, purified it, and sent it off to home brewers.[47] The following year, Clean Water Services officially started the Pure Water Brewing Challenge. Media outlets responded. Clean Water's Mark Jockers, the government and public affairs director who spearheaded the first Pure Water Brew Challenge, noted that the 2015 challenge resulted "in 400 stories worldwide."[48] National Public Radio, the *Guardian*, and *Food & Wine* all covered the effort.[49] Clean Water Services successfully demonstrated that going beyond a simple education campaign opened new avenues for talking about water recycling.

The Clean Water Services 2014 and 2015 competitions demonstrated the feasibility of partnering with brewers to expand the conversation inside and outside of professional water circles around DPR. In the years since, a small but notable number of brewing competitions that use recycled water have emerged with the goal of breaking down the yuck factor barrier across multiple ranges of expertise. Initially, competitions aimed to engage policy makers and other water professionals such as the Water Environment Federation's Technical Exhibition and Conference's, "Sustainable Beer Smackdown" held 2015–17.[50] Similarly, in 2017 utility and reuse providers from Pima County, Arizona, partnered with researchers at the University of Arizona (Tucson), to develop a mobile, full-scale DPR demonstration trailer. The trailer not only traveled throughout the state demonstrating how DPR worked, it directly delivered DPR water to breweries participating in the 2017 AZ Pure Water Brew Challenge.[51] Like the Sustainable Beer Smackdown, the AZ Pure Water Brew Challenge took place in the exhibition hall of a professional conference center, a space that while ostensibly accessible to the public primarily engaged policy makers (e.g., from DEQs), professionals from public and private utilities, and the representatives of supporting businesses.

Earlier competitions, like the 2017 event, did result in breweries being able to sell excess beer to consumers. However, the Scottsdale event was the first competition to be widely open to the public. One brewer interviewed had participated in both the 2017 and 2019 events. When asked whether he saw similar tasting patterns, he responded, "No, not really," pointing to the difference between a conference attendee and someone at a festival. "You don't go to a trade show 10 states away that's about water quality and water treatment unless you are—you don't wake up one morning and go 'Oh, hey, that sounds like a really fun vacation.' No, you've got a PhD in something that sends you there," he mused. "Whereas walking along the canal in Scottsdale and watching [a] dragon shoot fire? Two totally different things. You know?"[52] In designing the One Water Brewing Showcase to engage a public rather than professional audience, Scottsdale Water shifted the scale of the conversation about water reuse. Arizona DEQ's willingness to work alongside the utility in developing a permit demonstrated the extent to which previous efforts had helped move the regulatory needle toward an openness to DPR. When asked "why do beer for Canal Convergence rather than just a water tasting," Scottsdale Water's then public information officer, Nicole Sherbert, pointed out, "I'm not going to get 30,000 people to show up to drink water. But I can get 30,000 people to show up to drink beer."[53] Scottsdale Water's public-facing approach is catching on; more recently, beers brewed from recycled water have appeared in Berlin (2019), Calgary (2020), and Singapore (2022).[54] Good beer, especially good beer made from what was recently wastewater, makes for good press.

## PERSUADING THE PERSUADERS

Getting to good beer, however, requires persuading beer brewers that treated wastewater can make good beer. Both the 2017 and 2019 festivals entailed recruiting brewers to participate in an exclusive brewers' tour of either the demonstration treatment train housed in a semitruck trailer (2017) or Scottsdale's advanced water treatment facilities (2019) to learn about the purification process. Recruitment brochures for participants in the One Water Brewing Showcase noted, "You [brewers] understand the importance of water and the science behind and safety of purified water. You're the best ambassadors water could ask for!"[55] One brewer, who learned about the tour from the Arizona Brewer's Guild said, "I jumped on that. I could not wait to see the inside of a building I drove past every single day and that was probably really cool on the inside."[56] After participating in the tour, he immediately signed up when invited to participate in the One Water Brewing Showcase. Another brewer noted that touring the facility, seeing how the water was treated and controlled, convinced him that the water was "probably the cleanest water on earth." He expressed frustration that regulations to date had limited the use of recycled water to golf courses.

"I think it's a shame," he told us. "Most of the water you buy off the shelf if you go buy some big brands, literally it's just tap water that's been treated and probably not treated as well as the recycled water we were getting."[57] After the tour and his experience brewing with DPR water, his response was exactly what a marketer would want: "I would never even think twice about using it to make beer [on a] full time basis at all."[58] The technological presentation did more than persuade brewers to participate in the brewing showcase. It invited brewers to imagine a future in which DPR water was already part of the municipal water infrastructure, and in the process, would better serve their needs. By speaking to the technical and chemical interests of brewers who participated, the tour demonstrated that DPR water would be a good building block for making good beer.

The organizing teams of the 2017 and 2019 brew fests coupled the opportunity to physically *see* DPR in action with chemical analysis reports. The reports offered their own kind of proof: "Most of the things [on the sheet] are zeros. . . . [They're] pretty much giving us like, just dead clean water that even has less [chlorine, minerals/ dissolved solids than tap water]," one brewer said.[59] Similarly, as the owner of a microbrewery pointed out, "the water came in actually cleaner than the water we normally use."[60] For brewers, analysis reports proved DPR's molecular quality in a "language" they already used in managing their brewing operations.

The aesthetic characteristics of the delivered DPR water told their own story to brewers about the water quality. "When you get it [the water] right from the truck it tastes like nothing. It tastes like absolutely nothing," one brewer who participated in the 2019 One Water Brewing Showcase told us.[61] A brewing team who participated in the 2017 AZ Pure Water Brew Challenge recalled pulling a sample glass of water from the tank during delivery. "This was pre-COVID," one of the brewers noted, "so we just stood around and passed this glass around, everybody drinking the same water. We don't do a whole lot of sensory on our water typically, but everybody was like . . . 'wow, this is water.'" The communal tasting of the delivered water reinforced producers' technological claims about the purity and quality of DPR water. The water's visible clarity further highlighted the quality: "It would have been a really good picture because it was just so crystal clear. Like, not what people associate with reclaimed water, at least, maybe what I didn't associate with reclaimed water."[62] Aesthetic characteristics, analysis sheets, and for those who attended, tours of demonstration facilities, combined to persuade brewers that they had received what water providers had promised: high-quality water ready for brewing with. All that remained was producing and sharing with the public the ingestible evidence in cups, cans, and bottles.

Making certain that consumers *knew* the beer was brewed with DPR water was central to the Arizona DEQ's restrictions. For those organizing Arizona brewers as DPR water ambassadors, the requirement to tell people the beer contained DPR water carried an added bonus: it meant that brewers were being invited as outreach collaborators and being pushed by regulatory restrictions to comply with that educational mission. The success of this carrot and stick approach relied on conceptions of DPR water's pureness. That pureness showed up charismatically in the row of zeros on the analysis sheets provided to brewers upon the water's delivery. As one brewer remarked: "We looked at the printout and it basically was $H_2O$, everything had been stripped down. There were no chemicals. There were no minerals. Everything was like point 000 parts per million. So, it was just straight water."[63] By keeping a laminated copy of the analysis sheet either at the bar or visibly posted, taproom staff were able to immediately provide information.[64] One brew team explained that "people might have been a little hesitant at first [about the beer], but if you had the conversation, if you had the minute or two to sit and explain it, how good the water was, how clean it was, the whole process of it, why we chose to do the project, it was pretty easy to convert people."[65]

It is notable that the public-facing competitions repeatedly mobilized adjectives like *clean* and *pure* in naming or describing their competitions. As food studies scholars point out, when food producers mobilize adjectives like clean or pure they are laying the groundwork for political arguments about how foods should be understood and produced, what people's tastes should be, as well as what people should be consuming.[66] Sometimes, as E. Melanie DuPuis notes in her history of milk, political arguments about cleanliness or purity promote technological fixes that skip the harder political work of bringing producers, distributors, and consumers together to combat a problem.[67] Framing DPR water as clean highlights the technical labor and regulatory care underlying DPR. Situating DPR water as pure does even more: it mobilizes an imagination of water as only $H_2O$.[68] In the process, other things present in water, even if they simply reflect the rocks and ground a water traveled through or over or the aquatic life that lives in water, become easier to understand as contaminants.

## BLANK SLATING THE ENVIRONMENT

Not all DPR processes use RO. The technology is expensive. Emerging research indicates that utilities could possibly sidestep RO treatment by treating with ozone and biofiltration to produce water that meets the criteria for potability.[69] However, procedures that don't use RO are more sensitive to levels of total dissolved solids (a measurement of the level of minerals in a water), temperature, and nitrogen levels such as what might be caused by farm runoff; if not appropriately diluted, the treated water can contain levels of nitrogen compounds or total dissolved solids that exceed regulatory standards.[70] Arizona's high average temperatures and elevated levels of naturally occurring minerals, as well as its significant levels of agriculture, make RO a logical choice for local proponents of DPR. RO offers the additional advantage of helping remove contaminants of emerging concern that are not yet—but may soon be—regulated. More critically, for proponents of DPR trying to work against the yuck factor, RO offers an end product that is, as one brewer pointed out, "absolutely the cleanest, crispest water you could ever imagine."[71]

From an engineering perspective, DPR water produced with RO makes really good water for beer brewers who like to control *all* aspects of their brewing. The "cleanest, crispest" nature of the DPR water delivered to brewers is actually *too* pure. Brewers "had to add in a lot of salts to counteract" the absence of minerals so that it could be suitable for brewing.[72] Salts are essential to yeast growth. But DPR water straight from the treatment train offered, as those analysis sheets reported, basically nothing. Due to the extreme level of purification, DPR water, unlike the current municipal water provided by cities like Scottsdale or Tucson, was a completely blank slate for brewers to build their brews from.

For brewers, this blank-slate nature of DPR was exciting. One brewer based in Tucson noted, "I'll try and strip the [municipal] water as clean as possible and then add those salts in to the ratios that I like for whichever style of beer I'm making."[73] Being able to start with water that is "just water," at least if water is defined as primarily $H_2O$ and very, very little of anything else, allowed participating brewers to use the same water source to create a wide range of different styles of beers. This includes beer styles traditionally linked to a specific geographic place with waters that distinctly reflect the earth—the terroir—they come from. The owner of a brewery who participated in the 2017 AZ Pure Water Brew Challenge pointed out, "They [the water producers] stripped everything out of it. Not just the bad stuff. The good stuff, too. . . . What we needed to do to that water to put it into a sense of bringing it back to something we could brew with." Her brewer pointed out that "the only thing we really did with it [the water], we just had to get the math right about how much to add back." For him, "getting the math right" allowed them to match their brewing water with the characteristics of water from Edinburgh, Scotland. That water, unlike the "clean" water delivered, is "pretty heavy on the bicarb and gypsum." As a result they adjusted the chemistry by adding magnesium sulfate and calcium chloride to get the water "as close to the Edinburgh services as possible."[74] The beer won the brewery a prestigious award in a subsequent competition.

Like water from the in-house RO systems used by many brewers with the capital and inclination toward brewing a range of beer styles, the DPR water delivered to brewers allowed them to precisely dial in the mineral makeup of the water used to reflect the naturally occurring levels of minerals from any desired (and chemically characterized) location. At the same time, the complete removal of characteristic minerals normally found in municipal water challenged brewers who were not already working with RO systems; it made obvious that for many small brewers, brewing is a constant give and take between the local environment shaping the characteristics of the water a municipality delivers and the beer producer. The variations brought on by that give and take can be valuable. But variations can also threaten the long-term viability of a business unless accompanied by a narrative that specifically valorizes perceptible differences.[75] When it comes to producing consistent beer, DPR's blank-slate nature could potentially level the playing field, at least in terms of water supply, between cash-strapped microbrewers and larger, more established breweries.

As discussed earlier, making foods with industrial terroir centers on minimizing and managing the perceptible traces of the tastes of place. DPR takes this premise further: the type of DPR water Scottsdale Water delivered to brewers carries the promise to transform the perceptible and imperceptible cues of a specific time and place—the wastewaters of what Kim Fortun terms "late industrialism"—into nothing.[76] The water produced lacks the molecular marks made by any place, plant, or animal life. It is a blank slate, ready to be reinscribed with whatever locally or nationally identified flavor profile a beer brewer or utility wishes to (or can afford to) recreate. DPR's industrial terroir, its ability to become whatever users need it to be, adds another layer of technological distance between everyday consumers and the systems and infrastructures shaping environmental quality. It says "everything is ok here. Put your attention elsewhere."

## REMAKING MEMORY

The central aim of blank slating water is not to convince beer brewers or industrial users of municipal water that technologies like DPR are a great fit for the region. Rather, it's to convince bodies in the here and now that their future sensing bodies will find DPR palatable as part of a sustainable approach to providing enough clean, safe water for all. Brewers found this an easy future to imagine. As one pointed out, "Water is the one thing that we have in plentiful supply until we don't. And you know, that's always in the back of our minds as brewers, because we think about that, we know how important it is to us."[77] He wasn't alone in his imagination of a water-scarce future. All of the brewers, brewery owners, and beer proponents interviewed understood that brewing beer in a water-stressed region like Arizona is perhaps a tenuous long-term proposition.

Proponents of DPR draw on a technocentric argument that "water should be judged by its quality, not its history."[78] Their argument falls within an integrated water management movement, summarized in the idea that "water in all its forms has value" and as such "should be managed in a sustainable, inclusive, integrated way."[79] Termed One Water, stakeholders within the movement aim to reorganize how regulations and government entities govern water. For example, rather than having one utility manage drinking water and another wastewater, a region adopting a One Water approach would integrate drinking water and wastewater management. The One Water approach is calling, in some senses, for a radical reorganization of deeply entrenched nineteenth- and twentieth-century ways of thinking about and managing water in the environment. In this framing wastewater is no longer seen as separate and in need of being directed away from a community. From green infrastructures that aim to allow water from storms to replenish underground aquifers rather than end up in sewers to the water recycling that defines potable reuse, people working in a One Water framework are trying to undo some of the consequences of nineteenth- and early twentieth-century water management and governance.

The yuck factor, with its visceral reactions, threatens the goal of reorganizing water management. For proponents of DPR, finding a way to invite the consuming public to go from "yuck" to "yum" is the water utility equivalent of making gold from dross in efforts to reorganize how water is managed.[80] Scottsdale Water's decision to share DPR water from its recently built DPR treatment demonstration facility with the public via locally brewed beer carries distinct similarities to the approach taken by Eau de Paris. In both cases, the marketing teams for each municipality invited folks in their city, through advertisements and other forms of outreach, to use their bodies to test the water and taste for themselves. There were, of course, distinct differences: anyone of any age could drink the offered, free glass of water from Eau de Paris representatives, while one had to be twenty-one or older and able to purchase a ticket to sample the beers at Scottsdale's One Water Brewing Showcase. The water offered by Eau de Paris volunteers was already available throughout the city, while the water used by brewers who participated in the 2019 One Water Brewing Showcase required a special permit from the state to deliver the water to folks outside of the demonstration facility. Despite these differences, both events used the theater of taste to activate what I and my collaborator Marisa Manheim have conceptualized as the yum factor.

In contrast to the yuck factor, the yum factor happens when something tastes good enough that one *wants* to consume it again. The yum factor is more than just a yummy taste; it connects the molecules that make up flavors with positive social, cultural, and aesthetic experiences to create new memories. Indeed, efforts to activate the yum factor rely heavily on not just a single moment of tasting, but also the context around tastings. Beer at art festivals or on scientific expo floors, as well as summertime water tastings in the park or even at the end of a plant tour, are all designed to engage drinkers in the process of creating new, positive associations between bodily experiences and municipal water that move beyond what researchers call "pre-cognitive affective reactions."[81] Efforts to activate the yum factor are grounded in water producers' logical understandings of the molecular content, and thus the quality, of their water. In this, efforts to activate the yum factor suggest organizers' anticipation that participants in the theater of taste will be able to set aside their hesitancies or suspicions around municipal water.

Despite the significant buy-in from brewers who participated in the Arizona brew fests and worked to activate the yum factor, hesitancy and suspicion still occasionally emerged to counteract the technological optimism and charisma of beer brewed with DPR. While participating brewers recalled embracing the overall project, not all of their customers were as easily persuaded. For example, one brew team recalled their malt vendor's reaction: "He poured [the beer], and he's like, 'oh, man it looks great,' and we're kind of talking about it a little bit. And, before he tried it, he was like, 'You did what?' And then he tried it. He was like, 'Wow, that's really clean. It's really good.' But then some other brewers came in, and he was like, 'Oh, you gotta try this poop beer!.'"[82] The malt vendor's comment highlights both a childish joy in the shock factor associated with feces and a desire to share the experience with others. It also, however, hints at the difficulty in relying on bodily experience to undo culturally situated hesitancies. One brewery owner reported that they had a regular customer who refused to drink the beer brewed with DPR: "He just said he just couldn't because of working in sewage [treatment] for 30 years."[83] Hesitancy, suspicion, and even the little moments of humor such as the "gotta try this poop beer" all point to how past experiences intersect with present and even future moments of sensing.

By asking everyday people to focus on the aesthetic qualities of their water as markers for the care and attention inherent in the industrial process of producing municipal water, water producers in Paris and Scottsdale invited nonexperts to support the utility in envisioning a positive, rather than negative, environmental future. For Paris, the positive environmental future imagined is one in which inhabitants choose municipal water over bottled, in which inhabitants will support the utility's efforts to maintain and update the system, all in the face of a changing climate. Scottsdale faces similar needs—that is, like all cities it wishes to reduce plastic usage and provide high-quality water to all inhabitants—but in a more water-stressed environment, although recent European droughts indicate potentially drier water futures for Paris, too. Indeed, Scottsdale recently informed an unincorporated suburb that starting in 2023, the city would no longer sell water for delivery outside of its city boundaries, in line with its drought management plan.[84] Over the past decade, an increasing number of articles in venues from the *New Yorker*, to the *Atlantic*, to the *National Review* have highlighted the dramatic impact of drought on the southwestern United States. All speculate about the region's viability. Abandoned boat ramps at Lake Mead and the white "bathtub ring" at Lake Powell (both sources of water for the Phoenix region) visibly testify to the water shortage. Doomsday, it seems, is practically here. Yet the future Scottsdale Water invites consumers to imagine is one in which water-intensive businesses like breweries can thrive. That imagined future exists not because of a miraculous streak of wet years, but because inhabitants decided to support the municipality and utility in taking the decision to adopt DPR, even though such an infrastructural trajectory had, until recently, been deemed an option of last resort.

Efforts to activate the yum factor, playful as they may be, are political acts embedded within larger processes of decision-making. By engaging inhabitants, policy makers, and members of the press in using their bodies to "taste" the future, proponents of DPR are asking different publics to actively support legislative, regulatory, and infrastructural changes to the status quo. As participants taste, they enter into a physiological invitation to rewrite the connections between taste and memory, to erase past concerns not just about the quality of a single glass of water, but also about the capacity of technologies, regulators, experts, and governments to provide *all* people with access to safe and good water.

## MISSING ALTERNATIVES IN THE POLITICS OF CLEAN AND PURE WATER

The blank-slated water used in brewing beer, or provided in tastings, offers an incomplete picture of what the actual on-the-ground situation will be. Once DPR has transitioned from its liminal state, water produced using the technique would be blended with treated water from the currently in-use supply, at least as long as some raw water remains available for use. In the future anticipated by DPR's proponents, consumers would receive water most likely characterized by an improvement in the overall perceived quality. This is especially true for communities where ground and surface waters have high levels of naturally occurring minerals and salts; blending with DPR water will decrease the mineral and salt content of delivered water. On the other hand, experts predict that continued drought will further degrade the quality of raw water coming into treatment facilities. As such, the addition of DPR may simply allow municipal water utilities to maintain an aesthetic status quo rather than see perceptible decreases in the quality of the water they can afford to produce.

The quasi-public accessibility of blank-slated water is a temporary anomaly. In the past only researchers or utility operatives working in testing facilities had access to it. In the future, blank-slated water will once again retreat behind the relatively closed doors of municipal water production.[85] So what does the temporary availability of blank-slated water signal? To think through this question, let's return to the idea of the inflection point. As I point out in the introduction, inflection points mark when a range of future possibilities begin to coalesce toward a future inevitability; they are the point when a specific future starts to come into being, rather than a range of futures. It took significant effort, over multiple decades, to reach the point where policy makers in Arizona and elsewhere agree that it is appropriate, and even desirable, to explore breaking down the carefully constructed barriers dividing drinking water from wastewater. The presence of blank-slated water demonstrates the success of those efforts. At the same time, the presence of blank-slated water highlights that despite the limited legitimacy given to consumers' aesthetic responses in shaping treatment decisions, proponents of DPR have had to partially step away from the technocratic approach built over the twentieth century and engage consumers using the same tools—the senses—that consumers themselves rely on every day. This current moment in which blank-slated water is available, even in its limited forms, thus marks an inflection point in the making.

DPR is just one of many possible new tools, knowledge structures, and networks of expertise that could help address the problems of adequate water quantity and quality. The various salts and organic compounds like pharmaceuticals and personal care products screened out by RO don't just disappear. They end up concentrated into new waste streams that themselves may pose significant environmental challenges. Research on desalination plants, which use similar techniques to those used in potable reuse projects, demonstrates that the sheer concentration of what remains, if dumped directly back into the environment, threatens plant and animal life as well as soil health.[86] As more wastewater is recycled, the amount of concentrated micropollutants that utilities must find a way to remove will increase.[87] Drought threatens to further magnify the environmental impact of those concentrated micropollutants on any downstream life.[88] Although research indicates that additional treatment such as electrooxidation (which uses an electrical current to form powerful free radicals that can break apart the bonds holding together different organic chemicals) can successfully break down the majority of the remaining micropollutants, such treatments take a lot of energy.[89] Furthermore, most of the treatments proposed for addressing the problem of concentrated waste streams from RO have only been proven at laboratory or pilot plant scale.[90] It remains unresolved whether such treatments, even if environmentally exigent, would be fiscally feasible for municipalities.

There are alternatives that partially resist the blank-slate nature of potable reuse systems that rely on RO. For example, the experimental eco-village in the Netherlands, Ecodorp Boekel, treats its wastewater with photobioreactors, systems that rely on light to allow microalgae to break down micropollutants and pharmaceuticals found in water.[91] For the algae-based systems to work in Ecodorp Boekel, the village's human inhabitants have had to change their habits. Toilets can no longer be cleaned with bleach because of the negative impact the chemical could have on the downstream organisms responsible for cleaning the water.[92] Similarly, a small number of ecologically minded sanitarians in the United States and elsewhere suggest cities approach the problem of inadequate water by stepping away from the water-intensive approach of flushing urine and feces down drains.[93] They instead propose use of composting toilets. Like the experimental eco-village in the Netherlands, a move toward composting toilets would require people to change their habits and their relationship with their immediate environment—and if they wish to produce good compost, with what they consume.

Despite the fact that alternatives exist, either in the form of different ways of doing DPR or in different ways of finding "new" water sources, such alternatives play a minor role in mainstream discussions of potable reuse. Indeed, the current approach being promoted by many proponents of DPR in the United States avoids asking individual inhabitants to change any behavior, although water managers would certainly *like* to see behavioral change. Rather, the current approach situates consumers as complacent or uninterested in making any significant changes.[94]

## (IM)PERCEPTIBLE FUTURES

The subtle irony of the AZ Pure Water Brew Challenge and the One Water Brewing Showcase is that the DPR water used to brew the beer was a one-off. Both had to obtain a special permit to take the water beyond its production facility. The beers were like a concept car: present, functional to some extent, but only representative of a future that *could* be. They existed because Arizona regulators allowed them to exist in a limited fashion, because different public and private organizations collaborated to build events at which the beer brewed from DPR water could be served, and because of the willingness of a group of brewers to try something new—itself the product of a lot of work on the part of organizers. Every member of the public who interacted with the DPR beers in bars or at Canal Convergence (or even the straight DPR water served at the 2022 One Water Brewing Showcase) interacted with a concept, a possibility of what the future could be.

Tasting the future offered by twenty-first-century industrial terroir is an exciting proposition. Despite the significant work done over the twentieth century by municipal waterworkers to make water "taste like water," potable tap water obdurately continues to taste and smell—not just of chlorine and the other perceptible markers of industrial terroir, but also of the algae, rocks, minerals, and so on that it has come into contact with. Many people do not appreciate those tastes, especially if the tastes are musty, earthy, or sulfurous.[95] DPR offers a perceptible argument of what could be. When people taste the water offered during tours of Scottsdale's demonstration DPR treatment train or drink a concept beer made with DPR water, their abilities to look, smell, and taste are being enrolled in the here and now to make a gustatory argument about the future.

Frankly, from a flavor perspective, for many people accustomed to the taste of bottled water or filtered tap water, the ingestible argument DPR presents is pretty persuasive. The green, musty flavors that plague so many water producers in the metropolitan Phoenix region are absent from the DPR water served directly from Scottsdale's DPR treatment train. With its lack of minerality, the DPR water tastes remarkably—or eerily—similar to many mainstream bottled water brands. Given that current proposals for integrating DPR into municipal water sources anticipate blending the purified effluent with treated water from the regular source, once regulatory bodies take the step of allowing DPR, in the near future water will still taste slightly of the rivers, lakes, canals, wells, and aquifers it traveled through. Just less so.

A few possible futures immediately emerge when one considers the gustatory landscape where DPR water is blended with the already existing supply. If the quality of incoming raw water stays similar to its current levels, then the number of people able to detect tastes or odors at the physiologic inflection point—the threshold of perception—will drop. Unwanted tastes and smells will interfere less often in the work of convincing rate payers and policy makers that municipal water production is worth investing in. As the perceptible quality increases, previously hesitant or suspicious consumers may regain trust in their municipal water and reduce their use of bottled water (assuming municipalities are able to successfully sever the deep-seated disgust DPR initially activates). In contrast, for those who continue to perceive differences, their embodied knowledge may find itself further devalued. Finally, if the quality of incoming raw water decreases (as is projected for water-insecure regions), blending of DPR water with the incoming supply may simply allow municipalities to maintain current aesthetic quality levels for a time.[96]

No matter which future emerges, as long as DPR delivers on its promise of technological mastery over waters' aesthetic qualities in addition to its ability to make safe water, the work of going from "yuck" to "yum" becomes almost completely concentrated on psychological inflection points. I see a hopeful possibility in that moment of tension: realizing that the water you might drink is also the water you, your neighbors, and local businesses dump down the drain opens the door for thinking about, and better protecting, the environment that water comes from. No longer is the pollution going into a water system removed from sight until some breach occurs. Rather, it is the intimate, embodied reality that the water coming out of a tap could also be the water rinsing shampoo out of hair, washing sunscreen off of faces, or even flushing kidneys and intestines. These systems connect individuals with the reality that they are playing a role in the larger systems that create the environment DPR water will come from.

In contrast, journalist Elizabeth Royte, in her 2008 piece on Orange County's Groundwater Replenishment System, posits a different possible future: "No one wants to think too hard about where our water comes from. It's more likely that the virtuosity of water technology will let polluters off the hook: why bother to reduce noxious discharges if the treatment plant can remove just about anything? The technology, far from making us aware of the consequences of our behavior, may give us license to continue doing what we've always done."[97] Royte's argument highlights the perils presented by industrial terroir: that the success of removing the perceptible markers of place risks further entrenching practices that produce environmental harms. If DPR moves from a liminal state to an accepted and well-entrenched, "invisible" infrastructure, the lack of perceptible markers may simply make it even easier for inhabitants to assume that "that's just how it is," rather than take action.

Taking meaningful action, as discard studies scholars Max Liboiron and Josh Lepawsky point out, requires more than focusing on individual behavior or even awareness about the impact individual choices have on the aquatic environment—although those still matter. Instead, taking meaningful action calls for looking at "systems of discard."[98] Indeed, DPR's ability to reduce the perceptible markers of the environment does not necessarily mean that the technology's use must inevitably lead to new forms of not-knowing. Instead, the technology's ability to interrupt narratives about cleanliness and purity offers an opportunity for reconsidering already existing systems. By linking wastewater and drinking water, as well as their perceptible and imperceptible characteristics, DPR opens the door for regional policy makers to alter assumptions about the role water and wastewater treatment should play in their region. For example, unlike the approach embraced by Chicago in the late nineteenth century of sending water downstream for others (human or not) to deal with, DPR's circularity means that regulators could require that the stuff removed from treated water remain obdurately present within the local system. Similarly, the measurable accumulation of pharmaceuticals, personal care products, and other contaminants of emerging concern in the discards produced by DPR could facilitate efforts to hold industrial actors across the globe responsible for remediating the pollutants their products bring into the region. Such efforts would inevitably interrupt the current status quo. They would require a drastic reimagining of local and regional economies.[99] The success of DPR's multilayered promise in promoting meaningful action depends entirely on the political will of policy makers, politicians, and voters.

# CONCLUSION

## Flavor Stories

IMAGINE FOR A MOMENT a table of friends out for dinner. The venue is nice, much nicer than the restaurant they normally gather at. Maybe it's January and the group has decided to embrace the "Dryuary" movement, skipping alcohol for the month. Or maybe the group has a member who doesn't drink alcohol at all, and they have agreed to make things easier by just not ordering any. No matter; when the sommelier brings the wine list, one member asks her what waters the restaurant has available. She smiles, handing the group a separate list accompanied by a heavy piece of paper printed with a wheel split into thirteen equally sized sections. Four are associated with taste, one with mouthfeel, and the remaining eight with odor. The group looks surprised. "It can be difficult to describe water" the sommelier remarks, "so this month we have teamed up with a noted scientific establishment to help you become better acquainted with the tastes and smells found in water."[1] The recipients look over the wheel. They examine the eight odor categories available and then discuss them. Consensus proves easy; only one category—fragrant—sounds like something they would want to drink.

While this imagined moment exists only in the realm of possibility, the "Proposed Flavor Wheel for Raw and Treated Drinking Waters" (discussed in chapter 3) and its subsequent iterations have circulated through the water production and treatment world since its introduction in 1987. Its graphic interface highlights a realm of taste and odor that many water drinkers still remain at least partially unaware of despite the well-publicized failures of aging water infrastructures: the world of water production, where water is transformed from its "raw" state into a potable foodstuff to be delivered via pipes, taps, or third-party bottlers. Although waterworkers initially proposed the "Flavor Wheel for Raw and Treated Drinking Waters" to help streamline their treatment processes, the wheel now can be found around the globe. Its updated version is known by professional associations simply as the "2016 Drinking Water Taste and Odor Wheel" (see figure 6 in chapter 3). While it may not be used everywhere, the "Taste and Odor Wheel" shaped how waterworkers in the United States and many other countries around the globe were trained in the late twentieth and early twenty-first centuries to communicate about the aesthetic characteristics of the raw and treated waters they encounter.

The people tasked with producing municipal water have faced wide variations in the raw water quality coming into their facilities. Sometimes natural events like wind, drought, or temperature changes negatively impacted the quality of raw water. Human activities such as steel production in the early twentieth century, oil refining, and the emergence of intensive agriculture after World War II contributed additional challenges. For water producers on limited budgets, learning to erase the perceptible markers of raw water's wide variations meant forming professional societies, drawing together innovations from other disciplines, developing the ability to reliably compare experiences addressing novel taste and odor challenges, and partnering with an increasingly international group of collaborators. As waterworkers learned to successfully minimize the tastes and smells picked up by raw water in the environment, as well as from treatment, they made a product increasingly marked by industrial terroir. Unlike foods such as fine wines or cheeses that highlight the tastes of the places they were made, municipal waters became waters that consumers anticipated should taste of almost nothing.[2] In the process, perceptible knowledge about the polluting processes happening upstream or in town became knowledge primarily available to experts, rather than the general public.

The water people drank in the early twentieth century tasted and looked distinctly different than it does now. In the decades since people in St. Louis thought it perfectly normal to receive brown water, or residents in New England cloudy water, many consumers in places like the United States and France have come to expect the tap water in their homes to be the same all the time and to viscerally anticipate that it will be relatively uniform whether they be in Phoenix, Chicago, or Paris, even if they intellectually recognize that such expectations are unrealistic.[3] Compared to the early twentieth century, the tastes, smells, colors, and textures in municipal waters in the United States and France (and many other places) *are* relatively uniform. When those flavors are not uniform, it's easy to be suspicious of municipal water's safety. The rise of readily available bottled waters, especially purified, bottled municipal or spring waters with their low minerality and gustatory rhetoric arguing that good water is unremarkable, makes it easy to opt out of the harder work of going beyond liking or disliking to value water based on accessibility, equity, and environmental impact.

## INGESTIBLE DEBATES

Over the past few years, I've been thinking a lot about the way tastes and smells—flavors—tell stories about the environment to sensing bodies.[4]

The sensory science processes used to measure physiologic inflection points play a critical role in how experts tasked with making and monitoring our industrial food supply think about sensing bodies, both in the present and the future. These processes have resulted in amazing new sensory experiences, as any fan of flavored chips or candies might point out. Synthetic flavors are, after all, the neon colors of the gustatory world. Yet at the same time, the processes that generate knowledge about the molecular world are also used to erase important information about the places perceptible molecules come from. More critically, those processes combine with dominant societal values in ways that can erase how individuals, regulatory bodies, and critically, corporations care for the environment. With this in mind, I am curious: What if everyday people also used flavor to tell stories about the future?

In asking this question, I start from the premise that flavor is indeed molecular. Once flavor is understood as molecular, as discussed in chapter 2, it becomes an interactive media technology open to experimentation and recombination. Flavor can be designed to communicate narratives that extend beyond pleasure to provide critical information to minds and bodies.[5] I think this aspect of flavor, that it is capable of activating our rational brains alongside our emotional, affective selves, makes it an especially powerful starting point for conversations about what the future could be like.[6]

Stories have narrative structures: beginnings, middles, challenges, resolutions—sometimes—and conclusions. Flavor profiles, as ADL consultants Stanley Cairncross and Loren Sjöström pointed out in their promotional brochures and presentations, similarly have narrative structures. Initial taste, textures, the amount of time a sensation endures, and aftertastes all communicate something to the person eating. These edible narratives are at the heart of contemporary food product development.

Building a flavor story about the future requires work. Try on for size this question that I am currently using in prototyping activities of the Flavor Stories project: "Tell me a story about a day in your future where your city implements Direct Potable Reuse."[7] When I do this exercise with my students, I ask them to do a few thought exercises to help them through the process of developing their flavor story. First, I ask them to decide what type of future they anticipate water recycling will bring to their town. Is it hoped for? Feared? Unexpected? Next, I ask them to identify the core emotional (affective) elements they want to communicate through their flavor story. They then spend about five minutes jotting out a short scenario from a day in the future they are imagining. Using insights from that short scenario, I invite them to identify flavors that go with the affective elements they've identified. Finally, I ask them to imagine they are designing a paleta (a Mexican gourmet popsicle, often made with chunks of fruit) to tell their story, and to write down a recipe for the paleta. Sometimes I bring in gourmet jelly beans or salt water taffy and have them assemble an edible prototype on the spot. While students, and I, find the process of telling stories with flavor tricky, it also opens up conversations. We laugh together, groan together in disgust, and lick our lips over imagined deliciousness as we explore the different ways we each envision DPR transitioning from its status as a liminal innovation into that of either a failed or an adopted technology diffusing through legislative bodies and water treatment facilities. Throughout it all, the affective responses that show up in the flavors we choose to include or exclude suggest the potentially emerging futures from the current inflection point in the making that is potable reuse.

Despite the trickiness of telling stories with flavors, I want to invite you, intrepid reader, to try it out anyway, to take part in an ongoing Sensory Labor(atory) experiment in an edible examination of what food and sensory science might possibly look like if done differently.[8] The Sensory Labor(atory) is a research creation effort I started in 2016 in collaboration with the undergraduate students at Harvey Mudd College taking my courses.[9] The "experiments" conducted in the Sensory

Labor(atory) are grounded in disrupting everyday practices of sensory labor.[10] Disruptions build on comparative historical approaches and ethnographic fieldwork by using techniques from performance art as well as critical and speculative design practices to breach or upend familiar everyday experiences.[11]

So how do you prototype a flavor story? You can start like I did with my first prototype and just write things out on paper. Yet like most things, this is an exercise that benefits from hands-on engagement. First, go to your local nationally supplied grocer and buy a variety of small snack-like things. You'll probably note as you shop that it's really easy to find certain types of textures or flavors, but that others are missing. Try going to a different type of grocer to fill in the gaps. By this point in your ingredient-gathering journey, one of the core arguments of the Sensory Labor(atory) project will be very clear: the industrial food system does a good job of shaping what can be perceived. My nearest grocer, for example, offers a wide range of sweet and a smaller range of salty snacks. Snacks that center bitterness or umami or use vegetables as their base are harder to find. To get those, I have to travel to an international grocer; the nearest one of significant size requires a twenty-minute drive, or a seventy-five-minute bike ride. Prototyping a flavor story dumps one directly into the contemporary manifestation of US food culture.

Flavor Stories work best when they circulate. The next step in prototyping a flavor story is to invite some friends, family members, colleagues, or neighbors over for a night of conversation and snacking. Set out the different ingredients on a table, along with little plates or cupcake liners. If you're in a region like Phoenix, consider using the sample question I proposed around water futures. If not, I suggest you work with your invitees to identify a critical issue facing your local city council. Maybe it's a proposal to extend public transit, or to build a new freeway, or to give a big business a major tax incentive to move its headquarters to your hometown. Reformulate the question using different time frames: five years for a near future, ten years for a mid-future, twenty-five for a more distant but nonetheless accessible future. And then invite everyone to make a flavor story using the snacks you bought. The final step in prototyping flavor stories is to circulate those stories. Pair people up in groups of two or three—if you're really interested in experimenting, I suggest trying to connect people who do not agree—and have them share their stories with each other. They may need to include instructions: eat this first, then that, and finally that. Sit back and watch. What discussions ensue?

At this point, the second premise of Flavor Stories emerges. The act of building a flavor story and sharing it with someone else, especially when coupled with the gustatory humility of eating the flavor stories built by others, activates cognitive *and* affective domains. Theories of commensality suggest that eating together creates bridges. Flavor stories, by activating different modes of communication, ask people to use their whole bodies to think alongside someone else. This in turn can activate true curiosity and, I hope, break down communication barriers that often appear when talking about divisive issues.

The Flavor Stories project is still an experiment in progress. I present it here as a speculative exercise—and as a disruption to the normal conclusive flow of academic writing—because using flavor to tell stories about the future calls attention to the various inflection points that shift the direction an individual or a society takes. Inflections, once perceived, invite action. As such, the practice of telling flavor stories connects the histories present in this book with still emerging but not yet established trends in the technological management of the ingestible environment writ large. Looking for inflection points helps mark otherwise unperceived action. As flavor stories accumulate around a subject, especially if repeated on a yearly basis, they offer to help people identify the impact of change over time. As an experiment, the Flavor Stories project rejects the premises of blank slate futuring. It instead insists that the bodies present in the here and now bring their own histories to whatever future will be built.

## CREATING FUTURE REALITIES

Inflection points mark power. This makes them not just objects of analysis, but also objects open for disruption. Recently, Marisa Manheim, a graduate student colleague of mine and member of the Sensory Labor(atory), toured the Scottsdale water treatment facility. The tour ended at the DPR demonstration treatment train, which itself culminates in a rather ordinary drinking fountain. The tour leader offered tour participants the opportunity to try the water. "They served us the water chilled," Marisa remarked, "and told us it was at the perfect temperature." After my volunteer work on the streets of Paris teaching people how to improve the taste of their water (including refrigeration), I wondered, the perfect temperature for what?

If, as Cynthia Selin notes, "technologies not only intervene in present realities, they also create future realities," then the perfect temperature for managing thresholds of perception in the here and now is also a technology actively working to allow future systems to be built, future consumers to continue our current patterns, and future beer brewers to make something that carries both a whiff of danger (sewer beer!) and a whiff of sustainability (from local sewer water!).[12] As water workers and scientists continue to research how to manage water's flavors, they are putting in place a range of new possible paths. Transforming nonquality water, as marked by the smelly or nonsmelly molecules and microbes it contains, into quality water is a form of storytelling. That is why boosters of recycled water, like the promoters of Paris's municipal water or the marketing team shepherding Nestle's different water brands, invite mouths and noses into the work of determining whether a water is safe and desirable. Chill water, and the physiologic inflection point—the sensory threshold—shifts. In response, the water tastes better. When the water tastes good, it becomes easier for waterworkers, public officials, and others tasked with producing municipal water to get inhabitants to support infrastructure improvement or expansion.

The ability of the treatment processes used in making DPR to produce water that has a pleasing flavor profile certainly offers some potential upsides. Scottsdale Water's then public information officer, Nicole Sherbert, noted that she'd love to do a public tasting event with DPR water: "We have hardness in our in our tap water. But you don't have [hardness with] Advanced Water Treatment [DPR] water because it's gone through reverse osmosis." As Sherbert imagined the tasting she pointed out, "I think it would be interesting. I think if people really went, 'Oh my gosh, Advanced Water Treatment water is so much better than the tap water or the bottle of water,' then, you know, we've proven that point."[13] Evaluated on taste alone, DPR water could easily put bottled water producers out of business.

At the same time, good tasting water risks completely "recasting" place. The municipal water coming out of taps in the Phoenix area rarely arrives at perfect temperatures—when summer is at its hottest, water from the cold setting on my kitchen sink can border on painfully hot. If I let that hot water sit uncovered until it reaches room temperature, sniff it, and then drink it, I catch the molecular reminders that my water supply is home to a range of aquatic life. The earthy/musty/moldy odors of methylisoborneol or green/potato/cork of geosmin peek through, despite the ongoing (and generally effective) efforts of the workers at the City of Phoenix Water Services Department who produce my water. I might not nominate that glass of water for an award. But in the moment when I taste my water, the pipes, canals, and reservoirs that connect metropolitan Phoenix to the largely disappeared Rio Salado, or the miles away Colorado, appear. In contrast, if successfully implemented, DPR water would offer at best hints of earth, must, mold, potato, cork, or green. It could become, like the Scottish ale brewed in Arizona, whatever water producers wanted it to be.

## RETHINKING WATERS' FLAVORS

That good tasting water future is exciting. I am concerned, however, that the erasure of flavors from water risks further cementing a politics in which those with access to good water are made even more ignorant of those without such access. Will the flavorless nothing offered by DPR further enable the "permission-to-pollute system" that waterworkers like Chicago's Baylis and others throughout the twentieth century railed against even as their work to improve water's aesthetic characteristics facilitated political inaction on pollution? Or can the flavorless nothing offered by DPR act as a kick-start to policy makers and voters to reshape how we engage with our own polluting actions and those of the industries in our region?

In a recent series of pilot water tasting workshops run as part of the Sensory Labor(atory), participants saw the perceptible qualities of their municipal water—the saltiness, hints of aquatic life, and chlorine—as an indicator of systems of water provisioning that have failed inhabitants in some way. They are right to be concerned, at least in the big picture, about failures of water provisioning systems. But I think an alternative story is present in those perceptible qualities: the saltiness and hints of aquatic life offer an intimate link with the place we live. That could be something to celebrate.

Rethinking water's tastes and smells in ways that would allow celebration is its own work of making a flavor story. Even as tastes and smells tell molecular stories in the here and now about our waters' pasts and presents, these sensations can reveal alternative stories about the future. With that in mind, I think it time to remove the Taste and Odor Wheel for Raw and Treated Drinking Water (or its more recent form, the Taste and Odor Wheel for Drinking Water) from the "artifacts of ignorance" box and instead turn it into a flavor story. Grab a pencil. Find a piece of scrap paper, or if you don't like drawing circles, I've provided a template in figure 7.

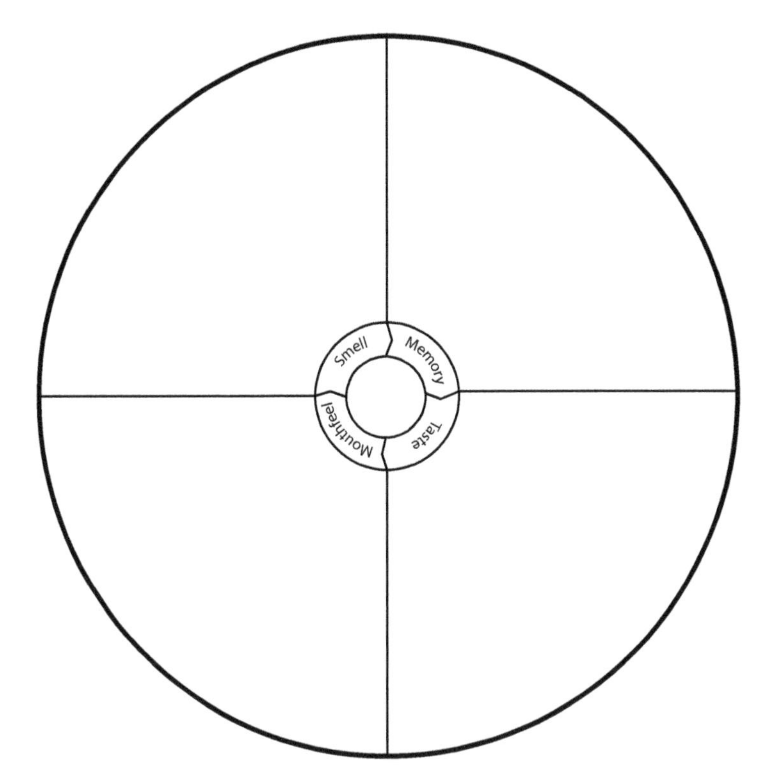


**Figure 7.** DIY flavor wheel for drinking water

Get a glass and start tasting. Don't forget to take a few field trips between tastings to gather water from neighboring communities. Make notes. Try to figure out where your water is coming from. Ask yourself: What and who lives in or near it? Whose lands and histories does it hold? As you take this journey, start filling in the blanks on the wheel. Modify it to account for things I left out. Build recycled water into it. Or not. And when you're done, feel free to invite your local councilperson or state lawmaker for a cold (or warm) glass and get to talking about what your Flavor Wheel of the Future for Raw and Treated Drinking Water reveals about what water should and could look, smell, and taste like, and the regulatory steps needed to help make that tasty future possible for everyone in your region.[14]

# Notes

## ACKNOWLEDGMENTS

1. Ballestero, *Future History of Water*; and Liboiron, *Pollution Is Colonialism*.

2. *We Are All Made of Stars* was part of the Best of QuiltCon 2022 exhibit; to see the quilt visit www.quiltallthethings.com/quilts-1

3. Gérald Caussé, "Our Earthly Stewardship" (sermon presented at 192nd Semi-Annual General Conference of the Church of Jesus Christ of Latter-day Saints, Salt Lake City, UT, 2022).

## INTRODUCTION

1. ABC7 Chicago Digital Team, "Chicago Launches 'Chicagwa' Campaign."

2. ComEd employee by day as of this writing, Thomas has successfully garnered a notable following on viral video site TikTok (@6figga_dilla). Reed, "TikTok's Dilla Is Educating Followers."

3. "Chicagwa Water."

4. Steven Shapin points out that folks working in the food and consumer goods industries have leaned into, rather than avoided, the work of quantifying subjective experience. This is further explored in the next chapter. Shapin, "Sciences of Subjectivity."

5. For more on the history of environment as a concept, see Benson, *Surroundings*.

6. As the notable uptick in anosmia (smell loss) during the COVID pandemic highlighted, sickness, as well as age, hormones, and genetics, can all impact whether someone can perceive a sensory cue.

7. Sound dampening provides an interesting contrast in that, at least in buildings, it is seen as a good rather than a danger. Thompson, *Soundscapes of Modernity*.

8. There is a growing body of political geography literature examining the creation of "hydrosocial territories." The concept of hydrosocial territories builds on a turn in political geography toward understanding territory as constructed through social interactions, political institutions, physical structures, and environments rather than as something shaped by national boundaries (reviewed in Hommes and Boelens, "Urbanizing Rural Waters"). Scholars point out that the building of hydraulic infrastructures transforms socially shared ideas (imaginaries) about how the world works into regulatory structures as well as tangible things such as pipes, buildings, and tunnels. Not everyone in a territory shares the same ideas, however. Hydrosocial territories are thus what emerges when contested imaginaries get defined by a subgroup of people in a specific, spatially bounded location, and then transformed into things and regulations. For example, the process of building projects to move water from rural to urban locations, at times from outside of a city's natural watershed, creates hydrosocial territories. The Colorado River Watershed offers an especially relevant example of a hydrosocial territory whose legal definition enacts and perpetuates inequitable access to water, especially for regions with junior water rights. Hommes and Boelens, "Urbanizing Rural Waters"; Swyngedouw, *Social Power and Urbanization of Water*; Swyngedouw and Boelens, "'And Not a Single Injustice Remains'"; Boelens et al., "Hydrosocial Territories"; Cantor, "Hydrosocial Hinterlands"; Summitt, Contested Waters; Kuhn and Fleck, Science Be Dammed; and Robison and Kenney, "Equity and the Colorado River Compact."

9. *Progress on Household Drinking Water.*

10. For example, Theresa Montoya points out in her examination of uranium contamination of water in the Rio Puerco that regulatory failures around routine radionuclide sampling have coupled with "state-imposed neglect expressed through sparse reporting" to create unsafe conditions on the Diné (Navajo) nation. Montoya, "Permeable," 140–41.

11. As James VanDerslice points out, water supply infrastructure is more than a physical infrastructure; it also includes the quality of raw water supplies, operational capacity, and regulatory structures. Nikhil Anand provides a useful lens for considering the political effects of expertise-produced ignorance from his studies of water provisioning in Mumbai. Anand situates infrastructural decay within the dispersed expert systems responsible for managing water provisioning and measuring leakage, noting that infrastructural decay is both produced by human action and "brought into being by the political effects of the technologies that constitute the city's infrastructure." Anand, "Leaky States"; and VanDerslice, "Drinking Water Infrastructure and Environmental Disparities."

12. The United Nations reports that the quality of global freshwater supplies is increasingly endangered, in part due to climate change, and in part due to human-caused degradation of water supplies. UNESCO World Water Assessment Programme, "World Water Development Report 2020"; and *Progress on Household Drinking Water.*

13. Community Water Systems are "public water systems that supply water to the same population year round" through "pipes or other constructed conveyances to at least 15 service connections." US EPA, Office of Water, "Information about Public Water Systems."

14. Masco, *Nuclear Borderlands*; Liboiron et al., "Abundance and Types of Plastic Pollution"; Snyder et al., "Pharmaceuticals and Endocrine Disruptors in Water"; US EPA, Office of Water, "Contaminants of Emerging Concern"; Corsi et al., "Fresh Look at Road Salt"; and Daugherty et al., "Impact of Point-of-Use Water Softening."

15. Allison Carruth and Robert Marzec, in their introduction to a special issue of *Public Culture* about the impact of environmental image making, argue that "forms of image making and visualizing have become naturalized and are shaping ways of seeing and also perceiving twenty-first-century ecological realities." Although this book is not concerned with image making, I see a notable

parallel with the naturalization of the olfactory and gustatory landscapes produced through municipal water treatment in places like the United States. Carruth and Marzec, "Environmental Visualization in the Anthropocene."

16. Mallevialle and Suffet, *Tastes and Odors in Drinking Water*, 2.

17. Mallevialle and Suffet, *Tastes and Odors in Drinking Water.*

18. Joy Parr's examination of the decision taken by the community of Walkerton, Ontario, to stop chlorinating their water supply demonstrates the complicated ways that sensorial perception of risk overlaps with public health conceptualizations of risk. Parr, *Sensing Changes*; and Parr, "Local Water Diversely Known."

19. Parrinello, Benson, and Graf von Hardenberg, "Estimated Truths."

20. Trubek, *Taste of Place*, xvi.

21. Parker notes that seventeenth-century conceptions of terroir differed from those of the nineteenth and twentieth centuries. The ability to detect the *goût de terroir* demonstrated connoisseurship: "Place-based eating for these [seventeenth-century] connoisseurs was just the *opposite* of how we currently think of the notion. It was not framed in terms of the complexity that terroir would add, such as the mineral fragrance that today's connoisseur might expect to smell in a wine from Chablis, but in terms of which place of origin possessed the most pure, healthful, and delicate produce. . . . The best terroir was a terroir that left no earthy flavor whatsoever, allowing the unmitigated essence of the food to show through." Guy, *When Champagne Became French*; and Parker, *Tasting French Terroir.*

22. Guy, *When Champagne Became French.*

23. Brad Weiss points out in his ethnography of the local pork industry in the Piedmont region of central North Carolina that framing taste as coming from a specific place is a political move that producers use to carve out space for alternatives to pressures from large-scale industrial agrobusiness. Amy Trubek, Jean-Pierre Lemasson, Kolleen Guy, and Sarah Bowen echo this, highlighting the fluidity of the concept through examination of how contemporary adoption of controlled designation of origin practices has reframed terroir. Harry West further nuances these discussions by noting the ways in which practices of making foods with a taste of place have accommodated changing modes of distribution that in turn impact tastes of place. In contrast, Sarah Besky's work on Darjeeling—an industrially produced tea that in 1999 was

granted geographic indication status—demonstrates that the concept of terroir can serve to repackage legacies of colonial industrial production into a form of heritage craft. Weiss, *Real Pigs*; Trubek, Guy, and Bowen, "Terroir"; Bowen, *Divided Spirits*; Lemasson and Trubek, "Terroir Products in North America"; West, "Thinking Like a Cheese"; Besky, "Labor of Terroir and Terroir of Labor"; and Besky, *Darjeeling Distinction*.

24. Spackman and Lahne, "Sensory Labor." The idea of sensory labor grew out of a multiyear collaboration that Lahne and I spearheaded. It is, however, an idea that we cannot solely claim; an ongoing collaboration with Ana Ulloa, Sarah Tracy, Nadia Berenstein, Ella Butler, and indirect collaboration with Hayden Kantor, Nicholas Shapiro, Jody Roberts, Luke Stark, Ai Hisano, Alexios Tsigkas, Anna Nguyn and audience participants in the Stop Making Sense symposium held March 10, 2017, at the then Chemical Heritage Foundation/now Science History Institute in Philadelphia, PA, contributed significantly to the development of this concept.

25. The relative "invisibility" of sensory labor situates it in conversation with theories about immaterial labor. Immaterial labor is the work that we might most closely recognize as producing immaterial goods; that is knowledge labor. Immaterial labor focuses in on the "autonomy and creativity of labor, and workers' power to bring about change." Gill and Pratt, "In the Social Factory?," 5. Lazzarato sees immaterial labor as having two components, the "informational content" found in a commodity (e.g., programming knowledge critical to development of new apps) and the "activity that produces the 'cultural content' of a commodity" Lazzarto, "Immaterial Labour," 133.

26. Anova, von Oertzen, and Sepkoski, "Introduction: Historicizing Big Data"; and Rosenberg, "Data before the Fact."

27. I am indebted to Ann-Sophie Barwich for providing a brief primer on the philosophical differences between sensing and perceiving, especially given that both shape the practice of science. A central characteristic of scientific work, Bruno Latour suggests, is "detaching, separating, preserving, classifying, and tagging" objects of scientific interest so that they can then be put back together and circulated according to a specific discipline's requirements. Latour, *Pandora's Hope*, 39.

28. Besky's examination of the "industrial ecologies" that go into a "nice cup of [black] tea" offers a useful parallel argument, highlighting the way

that creating quality that appears consistent happens in multiple spaces and at multiple temporal moments. Besky points out that the creation of quality includes the work of tea brokers, numerical descriptions at auctions, the historical development of blending practices, and scientists' attempts to identify the chemical constituencies of tea, as well as characteristics that increase yield. Through the work of these different actors, quality is closely, but nonlinearly, linked to economics. Besky's examination of the work that goes into making black tea have a consistent flavor profile despite the constant variation in tea leaves offers a useful counterpoint to the case of water: read through the framings offered in this book, the sensory labor that happens in the auction house and blending room is what makes possible the standardized qualities of black tea, and those standardized qualities in turn facilitate the creation of flavor profiles amenable to mass consumption rather than connoisseurship. Besky, *Tasting Qualities*.

29. The definitional work around agnotology emerged from a series of collaborative workshops held in 2003 and 2005 and resulted in an edited volume outlining where the field might go. Proctor, "Agnotology"; Proctor and Schiebinger, *Agnotology*. Since then, studies of absences, ignorance(s), and (the often grassroots) efforts to fill those gaps have expanded significantly both in STS and elsewhere; cf. Rappert and Bauchspies, "Introducing Absence." Anand, *Hydraulic City*; Kinchy, Parks, and Jalbert, "Fractured Knowledge"; and Smith, "Missing, Martyred and Disappeared."

30. Galison, "Removing Knowledge."

31. Schiebinger, "Feminist History of Colonial Science"; and Proctor, "Agnotology."

32. Dawson, *Undercurrents of Power*.

33. Thompson, *Soundscapes of Modernity*.

34. Frickel and Vincent, "Unintended Organization of Ignorance," 184.

35. Frickel and Vincent, "Unintended Organization of Ignorance."

36. Murphy, *Sick Building Syndrome*.

37. Liboiron, *Pollution Is Colonialism*, 46–77.

38. Liboiron's work is especially concerned with how a scientific theory can contribute to ongoing colonial relationships, specifically with regard to "assumptions of access by settler and colonial projects to Indigenous lands for settler and colonial goals." When one understands that science can function

this way, Liboiron points out, it becomes clear that pollution is a symptom of "the logics, mechanisms, and structures of colonialism." Liboiron, *Pollution Is Colonialism*, 5, 15, 51; see also 46–77.

39. Hobart, "At Home on the Mauna," 31.

40. For more on this concept, see Hoover, *River Is in Us*; Jackson, "Scents of Place"; and Shapiro, "Attuning to the Chemosphere."

41. This idea is reflected in a recent move to examining how settler colonial logics shape environmental care and is reflected in earlier work looking at the relationship between human infrastructures and nonhuman life. For more on this see Bahng, "Pacific Proving Grounds and Settler Environmentalism"; and Benson, "Generating Infrastructural Invisibility."

42. Although as Susanne Freidberg shows in her history of freshness, that lack of attention depends on where one is situated in the food system, as well as whether one is living through a technological or regulatory transition. Freidberg, *Fresh*.

43. There is a growing body of literature demonstrating that despite the absence of gustatory and olfactory archives, smells and tastes, as well as their social impacts and management, can be fruitful points of examination. Some notable examples include Classen, Howes, and Synott, *Aroma*; Corbin, *Foul and Fragrant*; Howes and Lalonde, "History of Sensibilities"; Kiechle, *Smell Detectives*; Kettler, *Smell of Slavery*; Tullett, *Smell in Eighteenth-Century England*; Smith, *Sensory History Manifesto*; Kafka, "Ingestion/Power Hungry"; Albala, *Eating Right in the Renaissance*. Smith provides a useful bibliography of nineteenth- and twentieth-century sensory histories. Smith, *A Sensory History Manifesto*. Within food studies, a subset of scholars have fruitfully used sensory experience as a lens for exploring industrialization; for example, DuPuis, *Nature's Perfect Food*; Bentley, *Inventing Baby Food*; Freidbert, *Fresh*; Hisano, *Visualizing Taste*; and Spary, *Feeding France*.

44. Croissant, "Agnotology"; and Freeman, "Perfume and Planes."

45. Croissant, "Agnotology."

46. After its initial growth in rural sociology, the lion's share of contemporary diffusion research has focused on the technology sector. However, significant work also examines the role of diffusion of innovation in the public sector. At the same time, certain arenas such as water technology continue to be overlooked, perhaps in part because of the significant infrastructural lock-in

seen by parts of the sector. Rogers, *Diffusion of Innovations*; De Vries, Bekkers, and Tummers, "Innovation in the Public Sector"; Meade and Islam, "Modelling and Forecasting Diffusion of Innovation"; O'Callaghan, Adapa, and Buisman, "Innovation Theories Applied to Water Technology"; Khandelwal et al., "Improved Cook-Stove Initiatives in India Failed"; and Parr, "What Makes Washday Less Blue?"

47. Rogers, *Diffusion of Innovations.*

48. Geroski, "Models of Technology Diffusion," 604; Valente and Rogers, "Diffusion of Innovations Paradigm."

49. Innovations are not particularly effective unless picked up and used by others. Sociologist Everett Rogers, through his study of adoption of innovations across a range of sectors, suggested that adoption and diffusion of innovation could better be understood as a "universal process of social change" rather than uniquely differing between cases and fields. Rogers argues that innovations diffuse most effectively when their promoters orient their research toward the social needs of a cultural group rather than the innovation itself. Rogers, *Diffusion of Innovations*, 11, 19–21. In this, Rogers's work aligns with more contemporary critiques of efforts to "educate away" misconceptions, as well as turns within the realms of sustainability, social justice, anticipatory governance, and others to center community voices in discussions around sustainability transitions. For more on the idea of a public deficit and subsequent efforts to erase it, see *Public Understanding of Science* 25, no. 4 (May 2016).

50. For an overview of this debate, see Korsmeyer, *Making Sense of Taste.*

51. Historians of psychology generally credit Gustav Fechner with founding psychophysics in the nineteenth century based on the research of his teacher, Ernst Weber. Islamic scholar Omar Khaleefa suggests the field was actually developed in the eleventh century by Ibn al-Haytham. Fretwell, *Sensory Experiments*; Gamble, "Applicability of Weber's Law to Smell"; Heidelberger, *Nature from Within*; Hui, *Psychophysical Ear*; Hyslop, "Psychophysics"; and Khaleefa, "Founder of Psychophysics and Experimental Psychology."

52. Heidelberger, *Nature from Within.*

53. Grove, "Navigating Strategic Inflection Points," 58.

54. Or in creating what Ballestero calls a "future history." Ballestero, *Future History of Water.*

55. Whipple, *Value of Pure Water*, 19.

56. For a comprehensive overview of the legal battles around chlorination, as well as its impacts, see McGuire, *Chlorine Revolution*.

57. McGuire, *Chlorine Revolution*, 81–82.

58. Illich, *H2O and the Waters of Forgetfulness*.

59. Donaldson, "Chlorination Tastes and Odors," 888–89.

60. For an overview of the challenges (and politics!) around naming and talking with others about smells or tastes, see Huisman and Majid, "Psycholinguistic Variables Matter in Odor Naming"; Spackman, "Ordering Volatile Openings"; Croijmans and Majid, "Not All Flavor Expertise Is Equal"; Alač, "Beyond Intersubjectivity I"; Heymann, "Personal History of Sensory Science"; and Muniesa and Trébuchet-Breitwiller, "Becoming a Measuring Instrument."

61. Spackman and Burlingame, "Sensory Politics."

62. "As humans, we live in environments, amid technologies, learning by doing. Our bodies are the instruments through which we become aware of the world beyond our skin, the archives in which we store that knowledge and the laboratories in which we retool our senses and practices to changing circumstances. Bodies, in these senses, are historically malleable and contextually specific. Our senses are the conduits through which knowledge of technology and the environment flow and, through retuning habit and reflex, the ways we habituate to our changing habitat." Parr, *Sensing Changes*.

63. Whelton and Dietrich, "Drinking Water Odorants"; and Spackman, "In Smell's Shadow."

64. I have purposely not placed an end date on the lead contamination crisis listed, as the impact of exposure to lead in developing bodies lasts a lifetime and reverberates out into family and community. Baehler et al., "Full Lead Service Line Replacement"; Roy and Edwards, "Preventing Lead in Drinking Water Crisis"; Corasaniti, Kilgannon, and Schwartz, "Tainted Water, Ignored Warnings"; Goldbaum, "'Tasting Funny for Years'"; and Carmichael and Boyer, "Health Impacts from Harmful Algae Blooms."

65. Gerkin et al., "Recent Smell Loss."

66. I've previously written about the experience of people in West Virginia continuing to smell the licorice-like presence of crude-MCHM long after instrumental measurements determined the chemical was no longer in

the system. Nick Shapiro offers a parallel case of embodied detection of formaldehyde, something he terms the chemical sublime. Spackman, "In Smell's Shadow"; and Shapiro, "Attuning to the Chemosphere."

67. The aquifer feeding the Magic Valley region he grew up in was popularly suspected to have been contaminated by years of radionuclide-containing wastewater disposal practices from the Idaho National Laboratory. As children we joked the water would make us glow at night. That seems less funny as an adult, especially in light of acknowledgment that contaminated water was directly injected into the aquifer. Bugger, "The INL and the Snake River."

68. I see the twentieth-century remaking of water production as sharing similarities with the development of water systems in the United States, the United Kingdom, and France during the urbanizing period between 1790 and 1860. This period of urbanization, historian Carl Smith argues, was shaped by an "infrastructure of ideas" about what it meant to belong, an emerging divide between "nature" and "culture," the role of urban spaces in health and well-being, and understandings of the obligations current choices would have on future inhabitants. While the details are different, the questions of who should pay for and have access to municipal water, whether municipal water is part of nature, the way water shapes health, and what impact current choices may have remain embedded (even if not overtly) in the development of sensory labor practices around managing municipal water's quality and (aesthetic) qualities. Smith, *City Water, City Life*; Gandy, *Fabric of Space*; and Goubert, *La Conquête de l'eau*. For a non-US case specifically focused on the role visual aesthetics played in promoting Mexico City's own "infrastructure of ideas," see Banister and Widdifield, "Debut of 'Modern Water'."

69. Melosi, *Sanitary City*. The EPA noted in 2007 that "8% of water systems in the US provide water to 82% of the population through large municipal systems." EPA, "Factoids."

70. Spackman, "Just Noticeable."

71. Interestingly, David and Richard Stradling argue that it was distance from, rather than proximity to, the Cuyahoga River that made the burning river so troubling. Stradling and Stradling, "Perceptions of the Burning River." Recently, Chanelle Dupuis has argued, building on Hsuan Hsu's work, that environmental pollution activates a form of slow smelling. Dupuis defines slow smelling as "a negative force that is permitting the fast speeds of

industrialization and environmental change to continue, allowing [olfactory] atmospheres to slowly shift . . . so that environmental malodors, pollutants, or industrial by-products slowly become stable, accepted scents in a particular space." Hsu points out that twentieth- century practices of what he terms differential deodorization—the uneven access to conditioned (purified/cooled/treated) air—has entrenched environmental injustices. Dupuis, "Smell of Water"; and Hsu, "Smell of Risk."

72. Hawkins, "Politics of Bottled Water."

73. Gleick, *Bottled and Sold.*

74. After Marisa Manheim and I finished presenting at the 2022 AZWater conference, a consultant came up to push back against our suggestion that teaching people to appreciate the place-based, seasonal nuances of municipal water could provide a useful entrance for municipal utilities to improve their relationships with their customer base. He told us that utilities facing a future defined by DPR *don't* want their customers to think about where the water came from. Instead, they want customers to have a deep-seated trust in the utilities and technologies that do the work of making water. His comments highlight a larger imagination I see as embedded in the work to erase perceptible markers of place: a desire to undo relationships with place and land so it can become anything. This carries notable parallels to the forms of violence enacted through colonial efforts to uproot indigenous peoples from their ancestral lands. Wendy Espeland's study of the successful resistance to building a dam on Yavapai land outside of Pheonix in the 1980s highlights the different rationalities that can undermine such imaginations. Espeland, *Struggle for Water.*

75. My nascent conception of a gustatory *terra nullius*, which I only gesture at here, has its seeds in learning about masking agents as a food science student at University of Illinois at Urbana Champaign. Conversations over the last decade with generous colleagues, specifically Hi'ilei Hobart and Max Liboiron, whose attention to the ways that dominant scientific thought displaces or erases indigenous peoples and their relationships to place and land, further inform my thinking here. Hobart, "At Home on the Mauna"; Hobart, *Cooling the Tropics*; and Liboiron, *Pollution Is Colonialism.* Cynthia Selin and Jathan Sadowski's conception of blank slate futuring allowed me to draw the line between masking agents and colonial erasures ("Against Blank Slate Futuring"). I look forward to further developing the concept in collaboration.

## CHAPTER 1: INDUSTRIAL TERROIR

1. Gerstein, "Chloro-Phenol Tastes." It is worth noting that later reports would disagree with Gerstein's analysis; Arthur Gorman points out that "the absorption of chlorine by industrial wastes and sewage was so rapid that this lone line of public health defense against water-borne infections was seriously threatened" ("Chicago's South District Water Filtration Project," 6).

2. Gorman, "Chicago's South District Water Filtration Project," 7.

3. Historian Bill Cronon points out that boosters saw Chicago's geographic situation on the southwestern shore of Lake Michigan, and its access to rivers flowing into the lake, as a core strength in their argument that the city should become the nation's crossroads. Cronon, *Nature's Metropolis*. For a more contemporary take on Chicago's environmental history that seeks to extend Cronon's analysis, see Keating et al., *City of Lake and Prairie*.

4. Hansen et al., "Tastes and Odors."

5. Steel, "By-Products from Industrial Wastes."

6. Warren, *American Steel Industry*; and Mayer, "Politics and Land Use."

7. Fales et al., "Progress Report on Industrial Wastes"; Tarr, "Industrial Wastes and Public Health"; Melosi, *Sanitary City*; and Tarr, "A 'Sink' for an Industrial Waste."

8. Osborn, "Phenol Tastes in Chlorinated Water."

9. Tisdale et al., "Cooperative State Control of Phenol Wastes."

10. Raab, "Troubles in the Minneapolis Water Supply," 431.

11. "William A. Evans."

12. Evans, "How to Keep Well."

13. Quoted from "Annuaire des eaux minérales de la France," 95.

14. Carl Smith provides a useful overview of the way that the transformation of "landscape into cityscape" reconstructed what was considered natural, healthy, and safe. Smith, *City Water, City Life*.

15. Besky, *Tasting Qualities*.

16. McGuire, *Chlorine Revolution*.

17. Cited in Warren and Bartow, "Taste and Odor in Chlorinated Water," 882. See also Melosi, *Sanitary City*, 95. Melosi notes that in 1916 a Milwaukee-based water operator shut off the chlorine to address consumer complaints, which caused "50,000–60,000 cases of gastroenteritis and between 400 and 500 cases of

typhoid resulting in forty to fifty deaths." Episodes like this became increasingly rare as water workers embraced their role as guardians of public health.

18. Filby, "Water Supplies of Florida."

19. Amsbary, "American Water Works Association."

20. Amsbary, "American Water Works Association," 1–4.

21. Amsbary, "American Water Works Association," 1–4.

22. Earl, "Minutes," 251.

23. Amsbary, "American Water Works Association."

24. Howson, "Responsibility for Potability of Water Delivered."

25. Shapiro-Shapin, "Filtering the City's Image"; Whipple, *Value of Pure Water*; and Richards and Ellms, "Coloring Matter of Natural Waters."

26. Warren, *American Steel Industry.*

27. Steel, "By-Products from Industrial Wastes."

28. Harlow, Powers, and Ehlers, "Phenolic Waste Treatment Plant."

29. Harlow, Powers, and Ehlers, "Phenolic Waste Treatment Plant."

30. Schneider, *Hybrid Nature.*

31. Adam Mack's delightfully pungent recounting of the way Chicagoans and visitors engaged with the city's river and canals offers an in-depth engagement of the visible and olfactory markers of pollution during the nineteenth and early twentieth centuries. Mack, *Sensing Chicago.*

32. Ericson, *Water Supply System of Chicago*; and Melosi, *Sanitary City*, 55.

33. The wastes fouling Chicago's waterways were first made famous by Upton Sinclair's muckracking novel in 1906. Sinclair, *Jungle.*

34. Ericson, *Quality Problem*, 10; and Department of Public Works, Bureau of Engineering, Division of Water Purification, *How Chicago Protects Its Water Supply.* For a more extensive history of this effort, see Cain, *Sanitation Strategy for a Lakefront Metropolis*; Melanie Kiechle provides a fascinating account of how politicians' olfactory experiences overrode chemical expertise in Chicago to impose a series of quick-fix solutions to address the stench caused by the city's industries (*Smell Detectives* 11–12, 139–40, 146, 148–60).

35. Joel Tarr and colleagues' survey of wastewater technology over the nineteenth century found that the engineering perspective, that "the power of streams should be utilized to its fullest for sewage disposal," prevailed through World War I. For more on the debates around who should be responsible for treating wastes, see Tarr et al., "Water and Wastes," 245.

36. Gorman, "Chicago's South District Water Filtration Project," 5–6.

37. Greeley, "Operation of the Water Filtration Plants," 443.

38. Colton estimates 59 tons/year phenol were produced in 1905; 111 in 1910; 78 in 1915; 134 in 1920; 136 in 1925; 127 in 1930; and 136 in 1940. Cyanide levels followed a similar trend during that time period, while oil levels peaked in 1920 at 11,176 tons/year, then steadily declined through 1930 to 2,794 before jumping again in 1940. Sulfuric acid levels followed a similar trend to oils ("Industrial Wastes in Southeast Chicago," 97).

39. Geographic indications, the legal structures used to officially mark terroir within international trade circuits, have been mobilized across the globe as part of larger development efforts. Sarah Bowen notes that despite the hopeful rhetoric surrounding the application of geographic indications in developing countries, her research demonstrates that the marketing of foods as having terroir doesn't necessarily result in benefits to producers. Bowen, *Divided Spirits*; Bowen, "Development from Within?"; and Trubek, Guy, and Bowen, "Terroir"; and Hughes, "Coffee and Chocolate." See also Ofstehage, "Construction of an Alternative Quinoa Economy"; and Biénabe and Marie-Vivien, "Institutionalizing Geographical Indications in Southern Countries."

40. The reversal of the Chicago River played a core role in helping create a perceptible rhetoric of Chicago's water as modern, alongside the city's pump houses and cribs. Cain, "Raising and Watering a City"; and Cain, *Sanitation Strategy*. For an overview of how water intake cribs functioned, see Burdick, "Water Works Intakes." For more on the work of abstracting and reducing water, see Linton, *What Is Water?* For more on the visual rhetorics and promises infrastructures hold, see Banister and Widdifield, "Debut of 'Modern Water' in Mexico City"; Larkin, "Politics and Poetics of Infrastructure"; and Anand, Gupta, and Appel, "Introduction."

41. I could potentially highlight the "merroir" of Lake Michigan, but have chosen to maintain the term terroir for its capacity to encompass human labor that seems to be missing from the currently nascent scholarship around the idea of merroir. As a concept, merroir seems to be a neologism coined in relationship to New England oyster harvests. McMurray, *Consider the Oyster*; Teigen de Master et al., "Terroir in Transition"; and Claus, "Beyond Merroir."

42. Warren, *American Steel Industry*, 140–44; and "Report of Investigation of Pollution of Lake Michigan."

43. Hansen et al., "Tastes and Odors," 1495.

44. Roman Garcier, writing about pollution in nineteenth-century France, makes the argument that the emerging sciences of water analysis, bacteriology, and hydrology set in place the playing field for defining which forms of pollution were dangerous, as well as shaping industrial responsibility. This, Garcier argues, resulted in legal structures that struggled to "accommodate polluting discharges as legal objects." Garcier, "Placing of Matter." Historian Craig Colton's examination of wastes discharged into Lake Michigan's Calumet Harbor reflects Garcier's argument ("Industrial Wastes in Southeast Chicago").

45. Phelps, "Chemical Measure of Stream Pollution"; Liboiron, *Pollution Is Colonialism*; and Hamlin, *Science of Impurity*.

46. Phelps, "Chemical Measure of Stream Pollution," 526.

47. Liboiron, *Pollution Is Colonialism*, 46–77.

48. Liboiron, *Pollution Is Colonialism*, 51.

49. Liboiron, *Pollution Is Colonialism*, 57.

50. Ericson, *Quality Problem*, 14.

51. Ericson, *Quality Problem*, 14–15.

52. Baylis and Ehrhart, "Taste and Odor Elimination," 641.

53. Gorman, "Chicago's South District Water Filtration Project," 9–10.

54. Gorman, "Chicago's South District Water Filtration Project," 5.

55. Burgess and Newcomb, *Census Data of the City of Chicago*.

56. For example, H. H. Gerstein lived in Hyde Park at 5484 S. Everett Ave. Doherty, "South Siders Owe This Stranger."

57. Meade, "Tastes Carried by Water Affect Cooking"; and Shapiro, *Perfection Salad*.

58. "Sprague Urges Water Filtering."

59. Evans, "Waste of Water Cuts Pressure"; Bebb, "Voice of the People"; and Ericson, *Quality Problem*.

60. Ericson, "Universal Metering of Water Supplies."

61. Chicago mayor William (Bill) Thompson, famously known for his mob ties, worked with local alderman to repeal the previous administration's requirement for universal metering, which had been reached as a compromise with the US government to allow additional withdrawal of water from Lake Michigan. "Water Meters," *Chicago Daily Tribune*; "U.S. Keeps Eye on Thompson's Proposal"; and Doherty, "Thompson Defies War Officials." For more on

Thompson's political career and the impact it had on Chicago's expansion, see Wendt and Kogan, *Big Bill of Chicago*.

62. Gorman, "Chicago's South District Water Filtration Project"; and Ericson, *Quality Problem*.

63. Ericson, *Quality Problem*.

64. Baylis, "Percolation and Runoff," 40, 42.

65. Baylis, "Treatment of Lake Michigan Water for Chicago," 710.

66. Baylis, "Treatment of Lake Michigan Water for Chicago," 711.

67. "Makes City's Water Taste Like a Spring," *Chicago Daily Tribune*.

68. Howson, "Responsibility for Potability of Water Delivered," 1147.

69. Howson, "Responsibility for Potability of Water Delivered," 1148; Gorman, "Chicago's South District Water Filtration Project," 11; and "Filtration Plants," *Chicago Daily Tribune*.

70. By 1942 the estimated cost had fallen to $23,800,000; the plant's completion was stalled until 1947 due to wartime shortages in building materials. Howson, "Responsibility for Potability of Water Delivered," 1147; and "South Side Fights Pollution Peril," *Chicago Daily Tribune*.

71. Ericson, *Quality Problem*, 17.

72. Ericson, *Quality Problem*, 16–17.

73. E. B., "Filtered Water."

74. Although the south side filtration plant did not offer residents a direct location upon which to recreate, its building did extend what is now known as Rainbow Beach farther out into the lake, increasing the amount of usable parkland. A second plant would be built just north of Navy Pier and placed into operation in 1964. Gayton and Gorman, "World's Largest Filtration Plant"; Gerstein, "Lake Michigan Pollution and Chicago's Supply"; Smith, "South Side Gets 125 Acre Park"; and "So. Side Groups Oppose Water Filter Project," *Chicago Daily Tribune*.

75. Baylis and Ehrhart, "Taste and Odor Elimination," 638.

76. Baylis and Ehrhart, "Taste and Odor Elimination," 638.

77. Kiechle's examination of the reaction of Chicago inhabitants and politicians to stench in the city during the nineteenth century offers an earlier, parallel case of attempting to control a negative sensory input, specifically the foul smells from industry, to enable business growth (*Smell Detectives*).

78. Baylis et al., "Further Observations."

79. Baylis and Ehrhart, "Taste and Odor Elimination," 636.

80. Whipple, "Observation of Odor," 587.

81. Baylis and Ehrhart, "Taste and Odor Elimination," 636.

82. A few examples are Norcom and Dodd, "Removal of Odor and Taste," 1414–37; Spalding, "Activated Char in Water Treatment," 646–48; Spaulding, Preammoniation; Hansen et al., "Tastes and Odors"; and Ellms and Lawrence, "Causes of Obnoxious Tastes and Odors," 463–73.

83. Hansen et al., "Tastes and Odors," 1503.

84. Spaulding's complaint may have been influenced by his desire to promote the new method he had developed. Despite that more cynical reading, the ready acceptance of Spaulding's method by those publishing in the *Journal* about taste and odor problems reflects a shared concern, as does the uptake of Baylis's subsequent refinements. Spaulding, "Quantitative Determination."

85. Spackman and Burlingame, "Sensory Politics."

86. Parker, *Tasting French Terroir*, 7–8.

87. Spaulding, "Quantitative Determination," 1038.

88. For more on this see Daston and Galison, *Objectivity*.

89. For a deeper examination of how quality problems intersected with the professionalizing identity of water workers like Spaulding, Baylis, and others, see Tarr, "'Sink' for an Industrial Waste," 170–75.

90. Hui, *Psychophysical Ear*.

91. Ortlieb, Kügle, and Carbon, "Fechner (1866)," 3.

92. Ortlieb, Kügle, and Carbon, "Fechner (1866)," 3.

93. Ortlieb, Kügle, and Carbon, "Fechner (1866)," 4.

94. Ortlieb, Kügle, and Carbon, "Fechner (1866)," 14.

95. Hui, *Psychophysical Ear*, 8.

96. Stratton, "Review *Applicability of Weber's Law*."

97. To determine whether Weber's law applied to smells, Gamble drew on an emerging field of research dedicated to theoretical and instrumental measurement of the physiology of smells grounded in psychophysics. Before Gamble's efforts, a significant portion of psychophysical research into olfaction had primarily taken place in the Utrecht in the lab of Hendrik Zwaardemaker. Barwich, *Smellosophy*, 39–40.

98. Spaulding, "Some Quantitative Odor Determinations," 1113.

99. Flentje and Gerstein, "Report of Committee," 1738.

100. Flentje and Gerstein, "Report of Committee," 1738.

101. Flentje and Gerstein, "Report of Committee," 1749.

102. Thomas, "Calculation of Threshold Odor."

103. Spaulding, "Some Quantitative Odor Determinations," 1039.

104. Gullans, "Procedure for Making Odor Determinations," 977.

105. Gullans, "Comparison of Odor Elimination Treatments," 64.

106. Billings et al., "Specifications and Tests for Activated Carbon," 1133–1224.

107. Billings et al., "Specifications and Tests for Activated Carbon," 1137.

108. Vaughn, "Laboratory Control and Operating Experiences," 2137.

109. It is notable that the research reported here was done in partnership with the chemical director of the International Filter Company; this may have contributed to the research design. Harrison and Behrman, "Granular Activated Carbon at Bay City."

110. TONs appear as a central part of reports from other water works operators around the country as well as sanitary chemists. Billings, "Taste and Odor Control at Dallas"; Helbig, "Comparison of Tests for Activated Carbons"; and Gibbons, "Water Pollution by Petroleum Oils."

111. Thomas, "Taste and Odor Control on Lake Michigan," 1883, 1886.

112. Spaulding makes this especially clear in the acknowledgments of his 1942 article: "It was hoped, originally, to improve the odor technic by cooperative efforts. . . . [T]his attempt was fruitless. . . . [E]ach operator must be the judge of his own technic since it avails but little to standardize all details except the most important one of the individual." Spaulding, "Threshold Odor Tests," 905.

## CHAPTER 2: MAKING FLAVOR MOLECULAR

1. Izaguirre et al., "Geosmin and 2-Methylisoborneol from Cyanobacteria," *A* 708–14; Michael McGuire, Skype interview with author, February 26, 2014; and Mitchell, "MWD Flooded with Complaints," sec. WS1.

2. Michael McGuire, Skype interview with author, February 26, 2014.

3. McGuire et al., "Closed-Loop Stripping."

4. "History: Western Municipal Water District, CA."

5. Streicher, "Colorado River Aqueduct System," 1223–26.

6. McGuire, Skype interview with author, February 26, 2014.

7. Spackman and Burlingame, "Sensory Politics," 350–71.

8. US EPA, Office of Policy, "History of the Clean Water Act."

9. Quoted in Mallevialle and Suffet, *Tastes and Odors in Drinking Water*, 103.

10. This randomness would later reveal itself as a result of the chemical reality of chlorination: chlorine levels decrease over time and distance from treatment. Thus, consumers closest to treatment and distribution plants did not notice the presence of the musty odors, as the high levels of chlorine disguised them. Consumers located farther away from the distribution point, however, received waters with much lower levels of chlorine and were able to detect the off flavors present in the water. It's not that the flavor-causing geosmin and methylisoborneol disappeared; instead they were masked.

11. McGuire et al., "Closed-Loop Stripping."

12. Mallevialle and Suffet, *Tastes and Odors in Drinking Water*, 103.

13. Krasner, McGuire, and Ferguson, "Tastes and Odors," 34–39.

14. Mel Suffet, Skype interview with author, February 20, 2014.

15. Michael McGuire, Skype interview with author, February 26, 2014.

16. While not directly explored here, the crossing of disciplinary boundaries that brought insights from food and sensory science into the work of managing municipal water reflects sociologic insights into how scientific disciplines and (sub)fields form. In the case highlighted here, the analytic approach employed by sociologist of science Nicholas C. Mullins in exploring the development of molecular biology could easily be applied: McGuire came to FPA through his doctoral mentor, Mel Suffet, who in turn learned about its existence from a food chemist colleague, Stan Seigel, who worked across the hall at Drexel. Additional connections emerge in the next chapter: Per-Edvin Persson's interest in flavor wheels caught Suffet's attention, and they recruited others to work internationally on the effort to develop shared language around tastes and smells found in water. Mullins, "Development of a Scientific Specialty," 51–82; and Irwin (Mel) Suffet, interview with author, February 20, 2014.

17. Baird, "Analytical Chemistry."

18. Knorr Cetina, *Epistemic Cultures*, 1.

19. In describing flavor as molecular I draw on the mid- to late twentieth-century understanding of bodies as composed of cell-signaling infrastructures that depend on interactions between molecules and proteins. As Hannah Landecker argues, molecular biology, and its reliance on computational theorizing

around signal processing, changed how scientists understand the body: instead of thinking of bodies as metabolizing (breaking down and incorporating) environmental inputs, molecularization functions by thinking about cell signals as inputs and receptors as detectors. Landecker, "Social as Signal in Chromatin," 79–99. While my own training as a molecular biologist and food chemist makes a conceptualizing flavor as molecular an obvious intellectual path, my ability to think through this has been deeply shaped by Sarah Tracy and Hannah Landecker's work, as well as by the larger scholarly conversation around biopolitics. Tracy points out that by the end of the late twentieth century, sensory experiences were understood as molecular experiences; at the same time they were simultaneously understood (by large food and flavor corporations) as affective experiences open to shaping. This, Tracy argues, is deeply political. Bruce Braun argues that molecularization changes the register of risk, focusing on how bodies are understood in "terms of a global economy of exchange and circulation." Tracy, "Delicious Molecules"; Braun, "Biopolitics and the Molecularization of Life," 7; Landecker, "Food as Exposure," 167–94; Rose, "Politics of Life Itself" ; and Rose, "Molecular Biopolitics," 3–29.

20. Specifically the work occurred through collaboration between the Quartermaster Food and Container Institute for the Armed Forces and the Quartermaster Research and Development Command US Army Quartermaster Corps.

21. Meiselman and Schutz, "History of Food Acceptance Research," 199–216; and Peryam, "Field Testing of Armed Forces Rations."

22. The larger problem, he later stated in his preface, is that of measurement. Peryam, Pilgrim, and Peterson, *Food Acceptance Testing Methodology.*

23. The textbooks are not necessarily wrong in their assessment, as two of the foundations of present-day sensory science, FPA and hedonic testing for liking, were introduced during the symposium. Stone, Bleibaum, and Thomas, "Introduction to Sensory Evaluation"; and Lawless and Heymann, *Sensory Evaluation of Food.*

24. Berenstein, "Designing Flavors for Mass Consumption."

25. Cairncross and Sjöström, "Flavor Profiles," 308–11; Tracy, "Delicious"; and Berenstein, "Flavor Added."

26. Rebecca Lemov, "'Hypothetical Machines," 401–11. For an expanded discussion of quantification of lived experience, see also Igo, *Averaged American*; and Bouk, *How Our Days Became Numbered.*

27. Arthur D. Little, Inc., "Flavor Profile Describes Food Flavors," 30–31.

28. Caul, "Profile Method of Flavor Analysis," 6.

29. Caul, "Profile Method of Flavor Analysis." For a more recent description of FPA, see Lawless and Heymann, *Sensory Evaluation of Food*.

30. Caul, "Profile Method of Flavor Analysis," 6.

31. FPA imagined observers in ways similar to how food producers imagined ingredients: as things that ideally could be interchanged. In practice, both recognized that not all bodies, nor all ingredients, could actually be interchanged; one person may not be as capable of working in a group as another. Similarly, peppermint from one location may taste and act noticeably different in a formulation than peppermint from another location. For a more contemporary exploration of interchangeability as a social value, see Emily Martin, *Flexible Bodies*.

32. Caul, "Profile Method of Flavor Analysis," 15.

33. Caul, "Profile Method of Flavor Analysis," 26. For more on the idea of making the body into a measuring instrument under twentieth-century scientific regimes, as well as the work that goes into making people into standardized observers, see Shapin, "Taste of Science," 436–60; Shapin, "Sciences of Subjectivity," 170–84; Muniesa and Trébuchet-Breitwiller, "Becoming a Measuring Instrument," 321–37; Lahne, "Sensory Science"; Spackman and Lahne, "Sensory Labor," 142–51; Lahne and Spackman, "Accounting for Taste"; and Berenstein, "Flavor Added."

34. Sensory science does not yet have a standardized history. The descriptions I use here draw from my own experience and training, but similar bounds on bodily expression can be found by reading old issues of *Food Technology*, the *Journal of Sensory Studies*, *Journal of Food Science*, and so on. Muniesa and Trébuchet-Breitwiller, "Becoming a Measuring Instrument" provides a contemporary example of remaking the body's tasting practices to be more like that of an instrument rather than a human. For a personal history, see Heymann, "Personal History of Sensory Science," 203–23. Nadia Berenstein provides a gripping overview of the industry's development as a whole in "Flavor Added." Those interested in further methodological reading should see Lawless and Heymann, *Sensory Evaluation of Food*.

35. Caul, "Profile Method of Flavor Analysis," 23.

36. Caul, "Profile Method of Flavor Analysis."

37. TONs, flavor profiles, and the *Atlas of Odor Characteristics* (discussed later in this chapter) all act as what Bruno Latour terms "immutable mobiles." For Latour, immutable mobiles are what emerge when people, be they botanists or biologists, take their findings and turn them into "inscriptions," forms of information that can travel via paper (or its electronic equivalent) to scientists unable to be in the laboratory, or field, when the information is produced. Latour, "Visualization and Cognition," 1–40; and Latour, *Pandora's Hope.*

38. Cairncross and Sjöstrom, "Flavor Profiles," 308.

39. Berenstein, "Designing Flavors for Mass Consumption."

40. It is clear from the report that ADL's employees knew the identify of many, but certainly not all, of the flavorful molecules present in ingredients used in building the toothpaste, as some were pure molecules (e.g., cinnamic aldehyde) and others were naturally occurring compounds or extracts (e.g., glycerin). Arthur D. Little, Inc., "Report on Flavor Studies of IPANA," 9.

41. Arthur D. Little, Inc., "Report on Flavor Studies of IPANA," 18.

42. Arthur D. Little, Inc., "Report on Flavor Studies of IPANA," 2.

43. Sjöström and Cairncross, "What Makes Flavor Leadership?," 56–58.

44. Arthur D. Little, Inc., "Report on Flavor Studies of IPANA," 5.

45. Baird, "Analytical Chemistry," 267–90.

46. This analytical move is indebted to Schummer, "Impact Instrumentation on Chemical Species Identity."

47. Evan Hepler-Smith provides a useful history of this process in his forthcoming book, tentatively entitled *Compound Words: Chemists, Information, and the Synthetic World.*

48. Holness, "Gas-Liquid Chromatography and the Perfumer," 370–97.

49. This is a superficial gloss of the history and function of gas chromatography and mass spectrometry; apart from Gerontas ("Creating New Technologists") and Berenstein ("Flavor Added"); little in-depth sociological examination of the nuances, failings, and co-construction of gas chromatography and mass spectrometry exists. Bartle and Myers, "History of Gas Chromatography," 547–57.

50. The end result of the increased granularity around olfactory experience contributed, to some extent, to the undermining of the human nose as a reliable instrument in a way that echoes the impact of the microscope on early modern practices of seeing highlighted by Gal and Chen-Morris. However,

unlike vision, olfaction has resisted instrumental taming; as I have highlighted elsewhere, smells continue to appear in the environment in novel, unexpected, and often human-caused ways. Their appearance, and the subsequent inability to quickly identify and remedy such olfactory experiences, reinforces lack of trust in technical experts among the nontechnocratic public. Gal and Chen-Morris, "Empiricism without the Senses," 121–47; and Spackman, "In Smell's Shadow."

51. Purnell, *Gas Chromatography.*

52. Holness, "Gas-Liquid Chromatography and the Perfumer," 371.

53. Holness, "Gas-Liquid Chromatography and the Perfumer," 371.

54. Jennings, Leonard, and Pangborn, "Volatiles Contributing to Flavor," 587.

55. Jennings, Leonard, and Pangborn, "Volatiles Contributing to Flavor," 587.

56. Jennings, Leonard, and Pangborn, "Volatiles Contributing to Flavor," 587.

57. Time, as well as exposure to light and packaging materials, further contributes to flavor problems.

58. Fuller, Steltenkamp, and Tisserand, "Gas Chromatograph with Human Sensor," 711–24.

59. Funding on the pear project was provided in part by Pear Zone No. 1 and the Public Health Service; Gerber Products Company provided support for a companion study. Jennings, Leonard, and Pangborn, "Volatiles Contributing to Flavor"; Heinz, Pangborn, and Jennings, "Pear Aroma."

60. Jennings, Leonard, and Pangborn, "Volatiles Contributing to Flavor," 588.

61. Heinz, Pangborn, and Jennings, "Pear Aroma," 758.

62. Stewart, "Quality of Poultry Meat and Eggs," 246–48.

63. Stewart, "Challenge in Flavor Research," 5.

64. Stewart, "Challenge in Flavor Research."

65. Wick, "Flavor Update," 46.

66. Fuller, Steltenkamp, and Tisserand, "Gas Chromatograph with Human Sensor."

67. Fuller, Steltenkamp, and Tisserand, "Gas Chromatograph with Human Sensor," 712.

68. Fuller, Steltenkamp, and Tisserand, "Gas Chromatograph with Human Sensor."

69. Kendall and Neilson, "Correlation of Odor Responses," 567–75.

70. Kendall and Neilson, "Correlation of Odor Responses," 568.

71. Different molecules can result in the same smells for someone; that is, more than one molecule smells to people like "green."

72. Malone and McFadden, "Special Identification Detectors."

73. Ettre, "Kovats Retention Index System," 31A–41A. I am indebted to Nadia Berenstein for making me aware of Ettre's work

74. Industry historians have attributed this to its perceived capability for "solving the problem of uniform expression of retention data" Budahegyi et al., "Twenty-Fifth Anniversary Retention Index System," 213–307.

75. Harper, "Some Chemicals Representing Particular Odour Qualities," 354.

76. Dravnieks and O'Donnell, "High-Resolution Headspace Analysis," 1049.

77. Pippen, "Poultry Flavor."

78. Dravnieks and O'Donnell, "High-Resolution Headspace Analysis."

79. Dravnieks and O'Donnell, "High-Resolution Headspace Analysis."

80. Dravnieks and O'Donnell, "High-Resolution Headspace Analysis."

81. Harper, "Some Chemicals Representing Particular Odour Qualities."

82. Dravnieks, "Odor Quality."

83. Dravnieks, "Odor Quality."

84. Dravnieks, "Odor Quality," 799.

85. Dravnieks, "Odor Quality," 801.

86. Morana Alač's recent ethnographic examination of psychophysics in the lab offers useful insights into some of the things absented from the descriptor lists, specifically the "plural" character of olfactory sensing. Specifically, Alač points out that cardinal ratings (such as the 0–5 participants were asked to attach to each descriptor) done in a psychophysics laboratory are accompanied by gestures, facial expressions, and other things that fail to appear in the report, or in aggregate. The language of smell, Alač argues, is composed of descriptor words and "movements of the body that render, while participating in, our experience of odorants." Alač, "Beyond Intersubjectivity I," 46; and Alač, "Beyond Intersubjectivity II," 474–502.

87. Sophia Roosth, "Of Foams and Formalisms," 4–16.

88. I draw here on Rose, "Molecular Biopolitics." Rose suggests that emergence of new technologies in the life sciences functions by seeking to "optimize the vital future by action in the vital present."

89. Knorr Cetina refers to objects like this as "malleable laboratory objects." *Epistemic Cultures*, 27.

90. In a presentation about taste and odor issues and their relation to irrigation waters in the Southwest presented at the 1953 AWWA annual meeting, the author argued, "It may, therefore, be concluded that irrigation return flows containing phosphate and inorganic nitrogen will increase the frequency of sieges of tastes and odors in surface waters." Silvey, "Relation of Irrigation Runoff," 1186.

91. Thaysen, "Origin of Taint in Fish I," 99–104.

92. Gerber and Lechevalier, "Geosmin, an Earthy-Smelling Substance," 935–38.

93. Medsker et al., "Odorous Compounds in Natural Waters," 476–77; and Rosen, Mashni, and Safferman, "Developments in Chemistry of Odour in Water," 106–19; this timeline is reviewed in Izaguirre et al., "Geosmin and 2-Methylisoborneol from Cyanobacteria," 708–14.

94. Jüttner and Watson, "Biochemical and Ecological Control of Geosmin," 395–4406.

95. Parinet, Rodriguez, and Sérodes, "Influence of Water Quality on Off-Flavour Compounds," 5847–56.

96. McGuire et al., "Closed-Loop Stripping." Introduced in 1973 by a Swiss chemist, the technique known as closed-loop stripping analysis takes advantage of the fact that smelly compounds easily evaporate into the air. As air circulates through the water sample, semi-volatile compounds (often the cause of odors) with lower evaporation points than water evaporate into the air stream run over the sample. A carbon filter then removes the compounds from the air, and one can then extract the compounds from the carbon using a solvent such as carbon disulfide. The extract can then be applied to a gas chromatography machine for analysis. Grob, "Organic Substances in Water, I," 255–73; Grob, Grob, and Grob, "Organic Substances in Potable Water, III," 299–315.

97. McGuire et al., "Closed-Loop Stripping."

98. McGuire et al., "Closed-Loop Stripping."

99. Suffet and Segall, "Detecting Taste and Odor in Water," 608.

100. Krasner, McGuire, and Ferguson, "Tastes and Odors"; and McGuire, Skype interview with author, February 26, 2014.

101. Arthur D. Little, Inc., "Food and Agribusiness Memorandum No. 2."

102. Arthur D. Little, Inc., "Food and Agribusiness Memorandum No. 2."

103. Arthur D. Little, Inc., "Food and Agribusiness Memorandum No. 7."

104. Michael McGuire, Skype interview with author, February 26, 2014; Krasner, McGuire, and Ferguson, "Tastes and Odors," 37.

105. Michael McGuire, Skype interview with author, February 26, 2014.

106. Krasner, McGuire, and Ferguson, "Tastes and Odors."

107. For other perspectives, see Persson, "Analysis of Two Muddy Odour Compounds," 1113–18; Piriou et al., "European Reassessment of MIB and Geosmin," 532–38; Worley, Dietrich, and Hoehn, "Dechlorination Techniques"; and Young et al., "Taste and Odour Threshold Concentrations," 331–40.

### CHAPTER 3: FUTURE SENSING BODIES

1. DuPuis, "Equipement: L'eau à Vau-l'Eau,"56–57.

2. Stewart. "Report From Europe."

3. DuPuis, "Equipement," 57.

4. Nathalie Jas outlines the impact of Carson's work on pesticide regulation in France, while Mark Stol gives a more global view. Jas, "Public Health and Pesticide Regulation"; and Stoll, "Rachel Carson's *Silent Spring*."

5. DuPuis, "Equipement," 57.

6. DuPuis, "Equipement," 57.

7. Chiu, "Water Law in France"; and Poncet, "La loi du 16 décembre 1964." Politician René Dumont, credited as the party's founder, famously signed off of television on April 19, 1974, by saying, "We will soon experience a water shortage, and that is why I am drinking a glass of precious water in front of you: because before the end of the century, if we continue our wasteful ways, water will be in short supply." Institut National de l'Audiovisuel, "René Dumont." For more on the rise of the Green Party, see Bess, "Ecology and Artifice."

8. Cartron, "De la sécheresse."

9. Cazaux, "Dégustateurs en blanc pour une eau pure."

10. Cazaux, "Dégustateurs en blanc pour une eau pure."

11. Butler, "Tasting Off-Flavors."

12. Butler, "Tasting Off-Flavors."

13. Hough, *Sensory Shelf Life of Food Products*. Olive oil presents an interesting case of this: instrumental measurements suggest that consumers should find stored olive oils acceptable for much longer than a sensory panel does. Lobo-Prieto et al., "Sensory Characteristics of Virgin Olive Oils."

14. West, "Thinking Like a Cheese."

15. I am grateful to Paolina Lu for taking the time to talk with me about the insides and outsides of this concept as I was developing it, and for her generous reading of this chapter.

16. The case of switching water supplies in Flint, Michigan offers an especially potent reminder of the dangers of thinking of water sources as interchangeable: the lower pH of the new supply from the Flint river resulted in lead leaching from pipes into the water system. Scholars who have examined the crisis point out that the failures extended well beyond some austerity-imposing policy maker's imagination of water as fungible. Katrinell M. Davis argues in her examination of the water crisis in Flint that the city's crisis was the "logical result of years of service disruptions, unchecked biases, and spatial inequalities," a view further supported by Laura Pulido's work. Tucson experienced a parallel, although lead-free, case in the 1990s. Davis, *Tainted Tap*, 6; Pulido, "Flint, Environmental Racism, and Racial Capitalism"; and McGuire and Pearthree, *Tucson Water Turnaround*.

17. Kenner, Mirzaei, and Spackman, "Breathing in the Anthropocene"; Lawless and Heymann, "Measurement of Sensory Thresholds"; and Phillips, "Taste Machine."

18. For an overview of the various tugs-of-war that have existed in this area, see King, "Aesthetic Attitude"; and Korsmeyer, *Making Sense of Taste*, 42. The eighteenth-century philosophical discussion of taste has been well documented and explored. Other notable works include Korsmeyer, *Aesthetics*; Howes and Lalonde, "History of Sensibilities"; and Shapin, "Sciences of Subjectivity."

19. The way one reacts to and judges the quality of objects is itself a marker of aesthetic taste. "Taste classifies, and it classifies the classifier," Pierre Bourdieu argues in his examination of French social class and ways of interacting with the world. For Bourdieu, the ability to identify what is aesthetically pleasing is informed by one's habitus—the world of people, objects, and actions in which one grows up and currently operates. Through an examination of the habitus of various social classes in France, Bourdieu concludes that only the bourgeoisie have achieved the mindset necessary for aesthetically experiencing the world as a disinterested observer. Bourdieu argues that French tastes for food are largely defined by class-based imaginations of how the body should function. Workers' bodies, he posits, are valued for their strength

rather than size, whereas the bourgeois body is valued for its ability to remain tightly controlled by the mind. Understood this way, one's taste for something reflects where one came from and what one has been taught to value. Bourdieu, *Distinction.*

20. Korsmeyer, *Making Sense of Taste*, 42.

21. Saito, "Everyday Aesthetics," 89. I am drawing from Saito's book in highlighting the particular markers of what has been thought worthy of aesthetic attention.

22. Korsmeyer, *Savoring Disgust*, 8.

23. For more on the history gestured at in this paragraph, see Hamlin, *Science of Impurity*; Goubert, *La Conquête de l'eau*; Kiechle, *Smell Detectives*; Roberts, "Death of the Sensuous Chemist"; Melosi, *Sanitary City*; Smith, *City Water*; and McGuire, *Chlorine Revolution.* The argument about the link between making microbes visible and the rise of mass consciousness around hidden dangers is drawn from chapter 2 of Spackman, "Transforming Taste."

24. Sayre, "French Connection."

25. Pezon, *Le service d'eau potable en France.*

26. See Schulhof, "Water Supply in the Paris Suburbs."

27. Pavenello, "International Association on Water Pollution Research"; and Per-Edvin Persson, personal communication with author, August 7, 2014.

28. For an overview of the political economy of water privatization, see Bakker, "Political Ecology of Water Privatization"; and Swyngedouw, "Dispossessing H2O."

29. Buller, "Privatization and Europeanization"; Fitch, "Liquidating the Public Sector?"; and Lorrain, "French Model of Urban Services."

30. Buller, "Privatization and Europeanization."

31. Paris did not regain municipal control of its water system until 2010. The private management of water continues to consolidate: In 2022 Veolia finalized a deal to purchase Suez, further consolidating the power of these groups to shape the sensory qualities of water on a global scale. Kartit, "Veolia's Nine-Month Revenue Jumps."

32. Sayre,

33. Sayre, "French Connection."

34. Auguste Bruchet, interview with author, October 7, 2013; and Lorrain, "La firme locale–globale."

35. Dennis, "Judges Pick Olympia, Wash."; Noss, "Distinguished Lecturer."

36. Bauby, "Water Globalization"; and Schneier-Madanes, *Globalized Water*, 56. The Atlanta contract, for example, came about through a partnership with US-based United Water Services.

37. Auguste Bruchet, interview with author, October 7, 2013.

38. Auguste Bruchet, interview with author, October 7, 2013.

39. The implementation of FPA at Philadelphia Water came about as part of the joint research project among Drexel, Lyonnaise, Philadelphia Water, the MWD of Southern California, and Philadelphia Suburban. Gary Burlingame of Philadelphia Water was tasked at the time with developing standard operating procedures; Irene Taylor led implementation and was the one who trained Khiari. Gary Burlingame, correspondence with author, December 15, 2022.

40. Djanette Khiari, interview with author, December 17, 2014. Khiari recently passed away. The water world has, as a result, lost a fantastic resource. Thank you, Djanette, for sharing your insights and generosity with me and so many others.

41. Khiari, interview with author, December 17, 2014.

42. Auguste Bruchet, interview with author, October 7, 2013.

43. Auguste Bruchet, interview with author, October 7, 2013; see also Bruchet et al., "Iodinated Haloform Medicinal Odor."

44. Mallevialle and Suffet, *Tastes and Odors in Drinking Water*, 103.

45. Auguste Bruchet, interview with author, October 7, 2013.

46. Djanette Khiari, interview with author, December 17, 2014.

47. Per-Edvin Persson, personal communication with author, August 7, 2014.

48. A second symposium was held in Japan in 1987, a third in the United States in 1991.

49. Per-Edvin Persson, personal communication with author, August 7, 2014.

50. Auguste Bruchet, interview with author, October 7, 2013.

51. Mallevialle and Suffet, *Tastes and Odors in Drinking Water*, 102.

52. Mallevialle and Suffet, *Tastes and Odors in Drinking Water*, 2.

53. Mallevialle and Suffet, *Tastes and Odors in Drinking Water*, 2.

54. Meilgaard, Reid, and Wyborski, "Reference Standards for Beer Flavor Terminology"; Langstaff and Lewis, "Mouthfeel of Beer—a Review"; and Noble et al., "Standardized System of Wine Aroma Terminology." Noble

copyrighted the term Wine Aroma Wheel, and it consistently appears capitalized in the literature.

55. Shapin, "Sciences of Subjectivity," 178. For an in-depth examination of the formation of one specific taste community, that of Comté cheese producers, see Shields-Argèles, "Cooperative Model of Tasting."

56. This critique would be further bolstered by scientists' eventual adaptation of techniques to disguise wine's appearance during tasting sessions by using opaque glasses, addition of dyes, or altered lighting. Morrot, Brochet, and Dubourdieu, "Color of Odors"; Pangborn, Berg, and Hansen, "Discrimination of Sweetness in Dry Table-Wine"; and Zellner and Kautz, "Color Affects Perceived Odor Intensity."

57. The public expansion of the Beer Flavor Wheel from being primarily used by producers to consumers is relatively recent; see Harbison, "Daily Infographic, Beer Edition."

58. Irwin (Mel) Suffet, interview with author, December 8, 2014.

59. Per-Edvin Persson, personal communication with author, July 8 2014.

60. Mallevialle and Suffet, *Tastes and Odors in Drinking Water*, 108.

61. Burlingame, Khiari, and Suffet, "Odor Reference Standards"; and Suffet et al., "Taste-and-Odour Wheel after 30 Years."

62. Khiari, *Distribution Generated Taste-and-Odor Phenomena*, xxxvi–xxxviii.

63. Mallevialle and Suffet, *Tastes and Odors in Drinking Water*, 130.

64. Pederson, Yorke, and Stamer, *Work Plan for Schuylkill River Basin*; Billen et al., "Water Quality in Seine River Estuary"; and Burlingame, Dann, and Brock, "Geosmin in Philadelphia's Water."

65. Sustainability scholars term this a "global metabolic transition," pointing out that more recent trends in "developing" countries are occurring in different ways than in countries that have more recently undergone this transition. Schaffartzik et al., "Global Metabolic Transition." The similarity of concerns caused by urbanization and industrial agriculture may possibly be traced by consulting opportunities in China and elsewhere that Suffet and others recently reported in conversation as having participated in. I am deeply indebted to one of my reviewers for asking me to clarify how congruencies between Paris and Philadelphia showed up as congruencies on the water wheel; their insights here were crucial to the argument presented. Thank you, whoever you are.

66. Auguste Bruchet, interview with author, October 7, 2013.

67. Auguste Bruchet, interview with author, October 7, 2013.

## CHAPTER 4: THEATERS OF TASTE FROM THE BOARDROOM TO THE STREET

1. I highlight the gender here because of the historically gendered labor of professional tasting sommelier work in the wine world. As Anna-Mari Almila notes, this trend has recently shifted, with increasing numbers of women being represented as wine professionals and receiving training in the field. Almila, "Wine, Women and Globalization."

2. "Fine Water Academy."

3. Mascha, *Fine Waters.*

4. The course was a collaboration between Amy Bentley and Christy Shields-Argèles; it drew widely on Shields-Argèles's research into how Comté producers mobilized terroir to help preserve traditional agricultural practices. For more, see Shields-Argèles, "Cooperative Model of Tasting."

5. Ahmed, "Orientations Matter"; Leysne, "Journeys through 'Ingestible Topography'"; and Shields-Argèles, "Cooperative Model of Tasting."

6. Teil and Hennion, "Discovering Quality or Performing Taste?," 27. Teil and Hennion do not disambiguate between what Korsmeyer terms taste and Taste; rather, their joint perspectives writing about wine tasting and music leads them to join the two and argue that the thing that matters is the way taste comes into being.

7. Teil and Hennion, "Discovering Quality or Performing Taste?," 28.

8. Muniesa and Trébuchet-Breitwiller, "Becoming a Measuring Instrument."

9. Spackman, Manheim, and Barua, "Tasting Water at Canal Convergence 2021."

10. Not all bottled waters are mineral water; in the United States the vast majority of bottled water available on grocery store shelves is purified water. For some in France, the distinction between purified bottled water and mineral water is distinct, accompanied by personalized choices of certain mineral waters for certain lifestyle needs— e.g., Vittel for sports hydration, Hépar for digestive distress. The mineral water offerings in the United States have recently increased, although the lack of a strong history of hydrological science means

the distinctions between purified and mineral water are less clear for many US consumers. No matter what, companies with bottled waters in their portfolios have taken advantage of bottled water's flexible meanings to build business. For more on the ascendance of bottled water, see Royte, *Bottlemania*; Gleick, *Bottled and Sold*; Hawkins, "Politics of Bottled Water"; Wilk, "Bottled Water"; and Marty, "La consommation des eaux embouteillées." For an overview of the French tradition of water cures and science, see Weisz, "Spas, Mineral Waters, and Hydrological Science"; Mackaman, *Leisure Settings*; Jennings, *Curing the Colonizers*; and Spackman, "Transforming Taste," ch. 1.

11. Walker, "Bottled Water Leaks Volume." Peter Gleick highlights a widely disseminated New York City blind taste test in 2005 that showed NYC-based drinkers in many cases actually preferred tap water over bottled. Gleick, *Bottled and Sold*. While as of the time of writing bottled water continues to have strong sales in the United States, younger consumers are increasingly eschewing bottled water in favor of refillable water bottles. Mintel, "Still and Sparkling Waters." Despite that trend, global bottled water consumption is estimated to have increased by 70 percent from 2010 to 2020. Bouhlel et al., *Global Bottled Water Industry*.

12. Elle, interview with author, October 8, 2013. The name is a pseudonym.

13. Nestlé Waters, "How to Taste Water."

14. The participants I talked to had differing memories of how many samples they had tried. One remembered over 100; the other remembered closer to 350.

15. For more on sensory science's efforts to transform subjectivity into objectivity, see Shapin, "Taste of Science"; Lahne, "Standard Sensations"; Lahne and Spackman, "Accounting for Taste"; and Phillips, "Taste Machine."

16. For an approachable overview of the physiology of olfactory experience, I highly recommend Barwich, *Smellosophy*.

17. Gilles, interview with author, August 19, 2013. The name is a pseudonym.

18. Elle, interview with author, October 8, 2013.

19. Gilles, interview with author, August 19, 2013.

20. Elle, interview with author, October 8, 2013.

21. Gilles, interview with author, August 19, 2013.

22. Gilles, interview with author, August 19, 2013.

23. Gilles, interview with author, August 19, 2013.

24. Gilles, interview with author, August 19, 2013.

25. Latour, *Pasteurization of France*, 85.

26. I was not privy to these meetings, so I cannot say for certain whether the samples were prepared to allow for an "unbiased" tasting—chlorine takes at least thirty minutes to evaporate from water, and the temperature a water is served at directly impacts the flavor detected. When questioned, Gilles could not recall how long the municipal water was allowed to breathe before serving, if at all, or at what temperature it was served. However, Elle's observation that the "normal" people at the expert tasting struggled to identify the difference between tap and bottled water, coupled with sensory studies that have found consumers are extremely sensitive to chlorine and use it to discriminate between tap and bottled waters, suggests that residual chlorine was probably present in the bottled tap water, facilitating discrimination. Cf. Teillet et al., "Preference of Bottled and Tap Water"; and Puget, "Chlorine Flavour Perception in Drinking Water."

27. As of 2019 the core idea of pour, observe, smell, taste remains on Nestlé's website. However, the space dedicated to the training has been collapsed down into a single page, with a 0:17 minute animated video replacing the more in-depth description found in 2014. Nestlé Waters, "How to Taste Water"; and Nestlé Waters, *L'Eau à la bouche*. By 2023, the directions for tasting had disappeared, replaced by narratives about healthy hydration, terroir, and sustainability.

28. Nestlé Waters, *L'Eau à la bouche*, 16.

29. The techniques Nestlé offered reflect those taught in *The Taste of Wine*, rather than the FPA approach used by water workers. Routinely credited with changing the way French and US drinkers taste wine, Peynaud's book explicitly defines the shape of the glass used, amount poured, and ingestion technique. It guides drinkers to examine the beverage before nosing the glass with gentle, controlled breathing, and to taste in a controlled enough fashion that one could easily spit out or swallow the beverage. Peynaud, *Taste of Wine*, 118–50.

30. Nestlé Waters, *L'Eau à la bouche*.

31. Shesgreen, "Wet Dogs and Gushing Oranges."

32. Nestlé Waters, *L'Eau à la bouche*, 12, 44.

33. Spackman, "Transforming Taste." For more on this, see Guy, *When Champagne Became French*; Trubek, *Taste of Place*; and Barham, "Translating Terroir."

34. See Lagard, Waks, and Alouf, *Realms of Water*. I see Nestlé's move to situate its waters as having terroir setting in place the possibility of a future argument to a not-yet-present set of strong regulatory regimes that its products should be exempted from environmental restrictions due to their unique characteristics. For more on the role of regulatory regimes in shaping the availability of bottled water, see Pacheco-Vega, "(Re)Theorizing the Politics of Bottled Water."

35. Elle, interview with author, October 8, 2013.

36. Elle, interview with author, October 8, 2013.

37. Elle, interview with author, October 8, 2013.

38. Elle, interview with author, October 8, 2013.

39. Gay Hawkins demonstrates the critical role that innovations in plastic production played in promoting the expansion of bottled water. Hawkins, "Commentary."

40. Notably, journals such as the *Journal of Wine Economics* regularly take up these questions. See Lecocq and Visser, "What Determines Wine Prices," in the inaugural issue.

41. Ironically, Thieblemont reports that the campaign was halted due to a lawsuit brought by the association of wine growers. Elisabeth Thieblemont, personal communication with author, February 27, 2023.

42. Elisabeth Thieblemont, interview with author, July 5 2013. Thieblemont changed positions in 2017; as of this writing she was working as the conseillère de la Direction générale en stratégie et prospective. As a result of the lawsuit, Eau de Paris was not able to permit reproduction of any of the materials from the campaign.

43. The Albien artesian well feeds the fountain in Lamartine Square derided by an earlier generation of tap water promoters (see the introduction to chapter 3).

44. Paris is credited with launching the first shot by introducing branded water carafes. See "La bataille marketing."

45. Lawless and Heymann, *Sensory Evaluation of Food*, 430–79.

46. Temporary market-style or park-style *buvettes* (bars) facilitated distribution. The buvette, a long table covered with a pink banner and set under a tent, sheltered volunteers, workers, and educational materials from the fickle Paris weather. Setup and breakdown at markets occurred in tandem with market vendors, giving early-rising and midday shoppers the opportunity to taste.

Park-style buvette setup and breakdown was timed to occur when crowds were anticipated to be the largest: midafternoon at the popular Abesses stop near Sacré-Coeur and all day at Gare de Lyon during the height of the summer travel season. The locations chosen not only situated tap water as a foodstuff, they also highlighted water's necessity.

47. Pamphlets were available in French at all booths, and at select booths they were also available in Chinese and English.

48. For an exploration of the idea of estimation, see Parrinello, Benson, and Graf von Hardenberg, "Estimated Truths."

49. At the time Eau de Paris put the labels on fountains throughout Paris, the European Union was still working through the process of developing standards for all member nations, with member states operating under an agreement from 1980. On December 16, 2020, the European Union adopted a revision of the 1980 Drinking Water Directive; the recast law came into force in January 2021. European Drinking Water Secretariat, "European Drinking Water."

50. See Whelton et al., "Minerals in Drinking Water."

51. Weisz, "Spas, Mineral Waters, and Hydrological Science."

52. For more on this process and its historical development, see Guy, *When Champagne Became French*.

53. Eau de Paris was not lying in placing the number at zero; like other micropollutants, pharmaceuticals and pesticides have very low detection limits, and when detected often fall below the threshold for reporting. For more on the politics of estimation, see Parrinello, Benson, and Graf von Hardenberg, "Estimated Truths."

54. One of the employees I interviewed noted that she would have preferred the utility talk about pharmaceuticals and pesticides.

55. Elisabeth Thieblemont, interview with author, July 5, 2013.

56. *OED Online*, "s.v. Valorization, n."

57. It is important to understand that this reincorporation of the utility into the public realm meant that quite a large number of technical employees were transitioned into public servants. Or, more plainly speaking, management changed, and the day-to-day employees did not.

58. I am indebted to the work of Max Liboiron for much of my understanding of how definition work functions in environmental controversies. Liboiron, "Redefining Pollution"; and Pine and Liboiron, "Politics of Measurement and Action."

59. Definition work in the food world intersects with what Xaq Frolich refers to as the "information infrastructures" of labeling laws. Frolich points out that nutrition labels materially organize food systems; similarly, food definitions act as information infrastructures that "standardize the relationship between [the food inside the package] and the consumer." Frolich, "Informational Turn in Food Politics," 146.

60. Ulloa, "Chef and the Flavorist," 188. Ulloa's conception of acuity resonates with geographical work on the role that visceral reactions can play in shaping political ideas. For example, Hayes-Conroy and Martin's examination of the Slow Food movement highlight that "how bodies come to *feel good* (or bad) in and through certain forms of acting" helps shape identification with a movement. Hayes-Conroy and Martin, "Mobilising Bodies," 271; and Hayes-Conroy and Hayes-Conroy, "Visceral Geographies."

61. For sensory professionals, taking part in blind tasting draws on practices of acuity. Ulloa, in comparing chefs and flavorists, suggests that sensory acuity depends on the ability to move beyond liking or disliking something and instead "make taste judgments about what is good and suitable." Doing so "involves technique, instrumentation, and, above all, shareability—that is, the ability to communicate a communal experience" ("Chef and the Flavorist," 188). Acuity, Ulloa argues, brings together "skill and knowledge, mind and body" (188); it is a form of sensory labor grounded in creating a community of shared practices of paying attention to flavor.

62. Pixelis, "Eau de Paris."

63. Pixelis, "Eau de Paris."

64. Euzen, "Voir, Goûter, Sentir."

65. Bénédicte Welté, interview with author, May 15, 2013.

66. As noted in the final chapter, the public engagement work my lab produces often uses tastings; see Spackman, Manheim, and Barua, "Tasting Water at Canal Convergence 2021."

## CHAPTER 5: ERASING PLACE

1. Scottsdale Water, "One Water Brewing Showcase."
2. Scottsdale Water, "One Water Brewing Showcase."

3. Baseline levels of water stress are calculated as a ratio of renewable annual water supply and annual water "withdrawals" that occur in a region. Hofste, Reig, and Schleifer, "17 Countries Face High Water Stress."

4. Fountain, "U.S. Declares Shortage on Colorado River"; and Hager, "Arizona, Nevada and Mexico See Water Cuts."

5. Allen, "Opinion"; and US EPA, Office of Water, "Contaminants of Emerging Concern."

6. A recent contemporary example of this would be the case of per- and polyfluoroalkyl substances (PFAS), a group of manufactured chemicals that emerged in the 1940s and exhibit significant persistence in the environment. The EPA only recently began work to regulate the presence of these chemicals in the environment. Public scholar Rebecca Altman's account provides an especially compelling insight into the very long lives of these chemicals. Altman, "20th-Century Synthetics Altered Us All"; US EPA,, "Preliminary Regulatory Determinations for Contaminants"; and US EPA, Office of Administration, "Basic Information on PFAS."

7. US EPA, Office of Administration, "List of Contaminants for Potential Regulatory Consideration"; and European Chemicals Agency, "Perfluoroalkyl Chemicals (PFASs)—ECHA."

8. Utility managers and others recognize that terminology matters when it comes to talking about DPR. The choice to use terms like *recycled* or *reclaimed* is political, linking DPR with the positive associations words like recycled and reclaimed hold. Beutler, "Your Potable Water Reuse Plan"; and Millian, Tennyson, and Shane, "Awareness and Acceptance Direct Potable Reuse." As discard studies scholars like Samantha MacBride and Max Liboiron argue, such framings overlook the industrial actors and regulatory systems responsible for pollution in the first place: "Recycling, as it exists today, does not in fact save ecosystems," MacBride notes. MacBride, "Does Recycling Conserve or Preserve Things?" For more, see MacBride, *Recycling Reconsidered*; and Liboiron, *Pollution Is Colonialism*.

9. Tenney, "Things to Know about Arizona's Drought."

10. Crewe, "Arizona Native Plants and Urban Challenge"; Konig, "Phoenix in the 1950s"; and DeJong, "Scheme to Rob Them of Land."

11. Conway, "Land Subsidence"; Slaff, "Land Subsidence and Earth Fissures"; and McGuire and Pearthree, *Tucson Water Turnaround*.

12. Sinking ground levels (subsidence), well failure, and development of large cracks in the earth's surface (fissures) encouraged Arizona's legislature to implement the Groundwater Management Act.

13. Carollo Engineers, Inc., "Long-Term Water Augmentation."

14. Not all communities equally took advantage of the resources found in wastewater. For a brief global overview of the different ways that communities used nightsoil, see Ferguson, "Nightsoil and the 'Great Divergence.'"

15. Pierre Goubert and Matthew Gandy provide useful entrances for considering how modern chemistry and germ theory shifted national conceptions of wastewater in Europe and North America, while Nancy Tomes and Alexander Parry offer insights into the ways that germ theory entered everyday practices in the home in the United States. Goubert, *La conquête de l'eau*; Gandy, "Rethinking Urban Space"; Tomes, *Gospel of Germs*; and Parry, "Delivering Bacteriology."

16. National Research Council, *Water Reuse*; and Rice, Wutich, and Westerhoff, "Wastewater Reuse across the U.S."

17. Gleich, "Scottsdale's Golf Courses in the Green."

18. City of Scottsdale, "Aquifer Storage and Recovery Wells."

19. Dow et al., "Sustainability of Indirect and Direct Reuse."

20. Scottsdale Arizona, "One Water Brewing Showcase."

21. Scottsdale Water, "One Water Brewing Showcase."

22. For example, Oregon's current rules state that "the use of recycled water for direct human consumption, regardless of the treatment class, is prohibited unless approved in writing by the Oregon Department of Human Services, and after public hearing, and it is so authorized by the Environmental Quality Commission." Arizona has a similar prohibition, although the current permit structure implemented in response to the 2017 and 2019 brewing challenges has set the stage for DPR adoption. Oregon Secretary of State, "Chapter 340, Division 55"; and Arizona Department of Environmental Quality, Arizona Administrative Code.

23. Arizona Department of Environmental Quality, "Draft Permit." Scottsdale Water played a core role in drafting the permit.

24. Arizona Department of Environmental Quality, "Draft Permit."

25. Arizona Department of Environmental Quality, "Draft Fact Sheet."

26. Waldby and Grendahl, "First Full Scale DPR Permit."

27. My research team and I came up against this limit while developing a public-facing water tasting exhibit in 2021; the limits meant that the utility could not "afford" to provide water for a comparative public tasting with an undetermined number of participants.

28. Licenses further shape how and when inhabitants in a region encounter DPR. Indeed, the current licensing structure of the Arizona Department of Health services is unable to license Scottsdale's Water Quality Lab because its tests for emerging contaminants do not yet fall under any regulatory structure. Waldby and Grendahl, "First Full Scale DPR Permit."

29. Scottsdale Arizona, "One Water Brewing Showcase."

30. It's notable that DPR is of great interest in the Netherlands, a region generally not described as water poor. Ibáñez Martín and Smits, "Living in Close Proximity to Wastewaters."

31. Tennyson, Millan, and Metz, "Getting Past the 'Yuck Factor'"; Ormerod, "Illuminating Elimination"; and Schmidt, "Yuck Factor When Disgust Meets Discovery."

32. Douglas, *Purity and Danger*; and Kristeva, *Powers of Horror.*

33. Wilk, "Bottled Water"; Hawkins, "Making Water into a Political Material"; Hawkins, "Politics of Bottled Water"; and Pacheco-Vega, "Human Right to Water."

34. Whipple, *Value of Pure Water*, 22.

35. Charles, *Suspicion.*

36. Charles, *Suspicion*, 5, 7, 6.

37. Charles does suggest that STS scholars, as well as critical feminists and social scientists, may find suspicion useful for theorizing "colonial modes of scientific knowledge and refusal of biomedicine in the anglophone Caribbean." Charles, *Suspicion*, 8.

38. Mertens, "Liminal Innovation Practices."

39. Scruggs, Pratesi, and Fleck, "Direct Potable Water Reuse."

40. Scruggs, Pratesi, and Fleck, "Direct Potable Water Reuse."

41. Rogers, *Diffusion of Innovations.*

42. For more on this, see the later section on remaking molecular memory.

43. For a more in-depth examination of the links between embodied cognition and sustainability transitions in water governance, see Manheim and Spackman, "Embodied Rationality."

44. Gatrell, Nemeth, and Yeager, "Sweetwater, Mountain Springs, and Great Lakes."

45. DeSalle and Tattersalle, *Natural History of Beer.*

46. For example, engineer Martin Brungard developed a program that helps brewers "dial in" the characteristics of the beer they'd like to make by "dialing in" the water profile they'd like to start with. See www.brunwater.com/about.

47. About the same time that Clean Water was inviting home brewers to use water containing 30 percent recycled water to brew, Milwaukee wastewater engineer Theera Ratarasarn was experimenting with using treated effluent to home brew. Ratarasarn personally collected treated effluent, further treated it, and then used the water to make a home brew that he named "Activated Sludge Wheat Ale." It scored well when evaluated by experts. Stratton-Childers, "More Beer, Less Yuck Factor"; and Flanigan, "Beer from MMSD Wastewater?"

48. Jockers attributes the idea to Clean Water advisory board member Art Larrance, who himself is credited with helping start the craft beer revolution in Oregon (and from there, the United States). Water Strategies, "Turning Reuse Water into Beer"; and Kopp, *Hoptopia*

49. Profita, "Why Dump Treated Wastewater?"; Associated Press in Portland, "Sewer Brewers"; and Pomranz, "Wastewater Beer's Sobering Moral."

50. Water Citizen, "Llllet's Get Ready to Reuuuuse!!!!"

51. "New Ways to Brew Reclaimed Water," *Pima County FYI External Newsletter.*

52. Interview with One Water Brewing Showcase Brewer #1 (OWBS1), July 15, 2020. The interview protocol used for interviews with brewers was codeveloped in collaboration with ASU sustainability doctoral student Marisa Manheim while Manheim was working as a member of the Sensory Labor(atory). Manheim led protocol development with my feedback and guidance. Interviews used were conducted by me and Manheim.

53. Nicole Sherbert, interview with author, September 28, 2020.

54. Some, such as Calgary's brew, were made in collaboration with membrane manufacturer Xylem. "Germany"; Jackson, "We Brewed Beer from Recycled Wastewater"; and Ratcliffe, "Singapore Craft Beer". The earlier approach used in Tucson and Oregon remains alive and well: in 2022, tech start-up Epic Cleantech partnered with a San Francisco brewery to make Epic OneWater

Brew; the beer was brewed using reclaimed wastewater from a luxury building for sampling at the 2022 Greenbuild Conference. Wells, "I Tried Beer."

55. Scottsdale Water, "One Water Brewing Showcase."

56. Interview with One Water Brewing Showcase Brewer #3 (OWBS3), July 16, 2020.

57. OWBS1, July 15, 2020.

58. OWBS1, July 15, 2020.

59. OWBS1, July 15, 2020.

60. Interview with One Water Brewing Showcase Brewer #2 (OWBS2), July 16, 2020.

61. OWBS1, July 15, 2020.

62. Interview with AZ Pure Water Brew Challenge Participant 5 (AZPWBC5), July 21, 2020.

63. Interview with AZ Pure Water Brew Challenge Brewer #2 (AZPWBC2), July 14, 2020.

64. OWBS2, July 16, 2020.

65. AZPWBC5, July 21, 2020.

66. Petrick, "Feeding the Masses"; Tompkins, "Sylvester Graham's Imperial Dietetics"; Bentley, *Inventing Baby Food*; Biltekoff, *Eating Right in America*; and Ray, "Back into the Sociology of Taste."

67. DuPuis, *Nature's Perfect Food*.

68. For more on how the idea of purity and cleanliness intersects with industrial food production, see Cohen, *Pure Adulteration*; and Smith-Howard, *Pure and Modern Milk*. Historian Christopher Hamlin and philosopher Hasok Chang provide extended investigation of the perils and opportunities present in situating water as $H_2O$. Hamlin, "'Waters' or 'Water'?"; and Chang, *Is Water H2O?*.

69. Roccaro, "Treatment Processes for Municipal Wastewater Reclamation."

70. Hooper et al., "Direct and Indirect Potable Water Reuse."

71. AZPWBC5, July 21, 2020.

72. AZPWBC5, July 21, 2020.

73. Interview with AZ Pure Water Brew Challenge Brewer #3 (AZPWBC3), July 8, 2020.

74. Interview with AZ Pure Water Brew Challenge Brewer #1 (AZPWBC1), July 9, 2020.

75. As Paxson and Weiss demonstrate, valorizing variation is central to the rise of "artisan" and "heritage" foods in the late twentieth century; such efforts distinctly sought to undo the early twentieth-century systems that saw variation as a lack of quality. Paxson, *Life of Cheese*; Weiss, *Real Pigs*; see also Petrick, "Feeding the Masses."

76. Fortun, "Ethnography in Late Industrialism."

77. OWBS3, July 16, 2020.

78. Ormerod, "Toilet Power."

79. US Water Alliance, "One Water Roadmap."

80. The process of wastewater reclamation could also be considered making gold from dross, in the sense of extracting rare earth minerals, biosolids that can be resold as fertilizers, and more.

81. Smith et al., "Public Responses to Water Reuse." For more on the cognitive, embodied reactions to DPR see Manheim and Spackman, "Embodied Rationality." I'm grateful to Marisa Manheim for sending this paper my way.

82. AZPWBC5, July 21, 2020. Both communications officers and brewers associated with the 2017 and 2019 brewing challenges noted concerns about news outlets using phrases like "toilet to tap" or "poop beer" in interviews.

83. OWBS2, July 16, 2020. Anecdotally, people involved in promoting DPR in Arizona have reported that those working in the water industry have been some of the most challenging people to get to try DPR.

84. "Colorado River Crisis Is Hitting Home."

85. I am not trying to say that municipal providers don't allow folks to come visit—they do! Rather, the number of folks able to take the time to come tour a municipal water production facility is small and is further circumscribed by limited personnel resources.

86. Elsaid et al., "Environmental Impact of Desalination Technologies."

87. Romeyn et al., "Contaminants of Emerging Concern."

88. Scholes et al., "Enabling Water Reuse."

89. Pérez-González et al., "State of the Art and Review."

90. Alshami et al., "RO System Scaling"; and Pérez-González et al., "State of the Art and Review."

91. Mohsenpour et al., "Integrating Micro-Algae into Wastewater Treatment"; and Smits and Martín, "Site for Multispecies Innovation."

92. Smits and Martín, "Site for Multispecies Innovation."

93. Kerri Jean Ormerod has done significant work in the Tucson, Arizona, area specifically, and in the desert Southwest more broadly, exploring attitudes toward water reuse. She points to eco-sanitarianism as a present but minor voice. Ormerod, "Toilet Power"; Ormerod, "Potable Water Recycling in Southwestern US"; and Langergraber and Muellegger, "Ecological Sanitation"

94. Charlotte Biltekoff and Julie Guthman draw on research into tech-based approaches to food innovation to demonstrate how imaginaries of consumers, when implemented by innovators, risk creating innovations that maintain the status quo rather than enabling actual change. Biltekoff and Guthman, "Conscious, Complacent, Fearful."

95. Participant observation at the public tastings I've done in collaboration with Marisa Manheim (2021/2022) and Shomit Barua (2021) with more than two thousand participants highlights the mixed reactions people have to perceptible cues. Some, such as the taste of "hose," regularly sparked happy recollections of childhood. In contrast, many participants responded with distaste to the other flavors found in the waters offered, especially the local water's saltiness or its "green" or "musty" flavors. Spackman, Manheim, and Barua, "Tasting Water at Canal Convergence 2021."

96. Dettinger, Udall, and Georgakakos, "Western Water and Climate Change"; Lin et al., "Synergistic Impacts of Nutrient Enrichment"; and Barbieri et al., "Climate Change Effect on Groundwater Quality."

97. Royte, "Tall, Cool Glass." I suspect that the increased cost of removing pollutants imposed by DPR, coupled with the increase in low-cost sensors, may instead activate new forms of environmental taxation not dissimilar from congestion pricing.

98. Liboiron and Lepawsky, *Discard Studies*, 6.

99. Liboiron and Lepawsky suggest that successfully addressing the systems responsible for discards requires interrupting taken-for-granted modes of perception (defamiliarization); demonstrating the historical, contingent contexts waste practices happen in (denaturalization); changing perspectives to view the peripheries created by dominant systems (decentering); and moving away from purity politics to allow difference (depurifying). Liboiron and Lepawsky, *Discard Studies*, 149–50. This final paragraph leans heavily on their approach and insights to suggest how one might take action.

## CONCLUSION: FLAVOR STORIES

1. This fictional scene is loosely inspired by a 2019 interview in *Penta* with water sommelier Jessica Altieri. Centrone, "20 Minutes with Jessica Altieri."

2. Or faintly of chlorine, depending on what one is used to.

3. Shapiro-Shapin, "Filtering the City's Image"; Whipple, *Value of Pure Water*; and Richards and Ellms, "Coloring Matter of Natural Waters."

4. I think there's an interesting parallel between water that tastes of nothing and other substances critical to survival, such as soil. These "all around, yet hardly apparent" substances require, Maria Puig de la Bellacasa suggests, "shifts in epistemological frameworks," that go beyond the rational into the affective realm. Puig de la Bellacasa, "Encountering Bioinfrastructure."

5. Aside from historical attempts at multisensory storytelling such as Smell-O-Vision, recent turns at the intersection of art, media, and engineering offer new insights into multisensory storytelling that include, rather than ignore, taste and smell. See Velasco, Tu, and Obrist, "Towards Multisensory Storytelling." As the work of environmental justice scholars highlights, there is a flipside to the playful nature of the narratives activated by multisensory encounters. For example, the concentrated, continued presence of unpleasant odors caused by a nearby refinery, hog farm, or similar facility can drastically rewrite the experience one has with a specific place. Jackson, "Scents of Place."

6. Smog meringues and trash popsicles are especially good examples of how flavor, ingested or imagined as ingestible, activates rational and affective engagement. Twilley, "Exploring Aeroir"; and Jennings, "Wastewater on a Stick."

7. Adjacent examples for thinking about climate change with the arts include Bentz, "Learning about Climate Change"; Sheppard et al., "Future Visioning of Local Climate Change"; Wang et al., "Data Edibilization"; and Forrest and Feder, *Climate Change Education*.

8. I've written about the ethical implications of sensory science's efforts to strip what I term "smell's shadow" away from sensory knowledge production and have offered up some initial thoughts on what doing sensory science otherwise could look like. Those shadows—the memories and experiences people bring to flavor—not only offer a challenge to objective production of scientific

knowledge about subjective experience, they also invite reconsidering the aims of sensory science entirely. Spackman, "In Smell's Shadow."

9. For more, see www.christyspackman.com.

10. Spackman and Lahne, "Sensory Labor."

11. I came to disruptions as an outsider to art/science, design (speculative or otherwise), and participatory engagement. My thinking and approach grew via generous conversations with Max Liboiron and Collin Bradford. My ideas have further been developed through reading and experiencing the work of a range of makers. The list is long. Here is where I started: Dunne and Raby, *Speculative Everything*; Voß and Guggenheim, "Making Taste Public"; Jickling and Reed, *Multiple Elementary*; Lindsay Kelley, "Tasting History"; Liboiron, *Founder/Worker*; and Raspet, "Formulation 0.10." The quiet insistence of Bree Goates that I had something to make added a necessary dose of confidence.

12. Selin, "Sociology of the Future," 1879.

13. Nicole Sherbert, interview with author, September 28, 2020.

14. This exercise is a variation on an activity collaboratively developed with doctoral students Marisa Manheim and Shomit Barua as part of our work building a public-facing water tasting exhibit for Scottsdale's 2021 Canal Convergence festival. Spackman, Manheim, and Barua, "Tasting Water at Canal Convergence 2021."

# Bibliography

ABC7 Chicago Digital Team. "Chicago Launches 'Chicagwa' Canned Drinking Water Campaign." *ABC7 Chicago*, May 3, 2022. https://abc7chicago.com/chicagwa-chicago-canned-drinking-water-lake-michigan-mayor-lori-lightfoot/11813944/.

Ahmed, Sara. "Orientations Matter." In *New Materialisms: Ontology, Agency, and Politics.*, edited by Diana H. Coole and Samantha Frost, 234–58. Durham, NC: Duke University Press, 2010.

Alač, Morana. "Beyond Intersubjectivity in Olfactory Psychophysics I: Troubles with the Subject." *Social Studies of Science* 50, no. 3 (June 1, 2020): 440–73. https://doi.org/10.1177/0306312720915645.

———. "Beyond Intersubjectivity in Olfactory Psychophysics II: Troubles with the Object." *Social Studies of Science* 50, no. 3 (June 2020): 474–502. https://doi.org/10.1177/0306312720915646.

Albala, Ken. *Eating Right in the Renaissance*. Berkeley: University of California Press, 2002.

Allen, Joseph G. "Opinion: These Toxic Chemicals Are Everywhere—Even in Your Body: And They Won't Ever Go Away." *Washington Post*, January 2, 2018. www.washingtonpost.com

/opinions/these-toxic-chemicals-are-everywhere-and-they-wont-ever-go-away/2018/01/02/82e7e48a-e4ee-11e7-a65d-1ac0fd7f097e_story.html.

Almila, Anna-Mari. "Wine, Women and Globalization: The Case of Female Sommeliers." In *The Globalization of Wine*, edited by David Inglis and Anna-Mari Almila, 191–212. London: Bloomsbury, 2020.

Alshami, Ali, Trevor Taylor, Nadhem Ismail, Chris Buelke, and Ligia Schultz. "RO System Scaling with Focus on the Concentrate Line: Current Challenges and Potential Solutions." *Desalination* 520 (December 15, 2021): 115370. https://doi.org/10.1016/j.desal.2021.115370.

Altman, Rebecca. "How 20th-Century Synthetics Altered the Very Fabric of Us All." Aeon, January 2, 2019. https://aeon.co/essays/how-20th-century-synthetics-altered-the-very-fabric-of-us-all.

Amsbary, Frank C., Jr. "The American Water Works Association 1881–1956." *Journal (American Water Works Association)* 48, no. 1 (January 1956): 1–4.

Anand, Nikhil. *Hydraulic City: Water and the Infrastructures of Citizenship in Mumbai*. Durham, NC: Duke University Press, 2017.

———. "Leaky States: Water Audits, Ignorance, and the Politics of Infrastructure." *Public Culture* 27, no. 2 (76) (May 1, 2015): 305–30. https://doi.org/10.1215/08992363-2841880.

Anand, Nikhil, Akhil Gupta, and Hannah Appel. "Introduction: Temporality, Politics, and the Promise of Infrastructure." In *The Promise of Infrastructure*. Durham, NC: Duke University Press, 2018.

"Annuaire des eaux minérales de la France." In *Traité de l'épuration des eaux naturelles et industrielles*. Paris: Librairie Polytechnique, Baudry et Cie, Editeurs, 1893.

Anova, Elena, Christine von Oertzen, and David Sepkoski. "Introduction: Historicizing Big Data." *Osiris* 32, no. 1 (2017): 1–17.

Arizona Department of Environmental Quality. Arizona Administrative Code, Title 18, Ch. 9, Article 7, Water Pollution Control § R18-9-707 (2016). https://static.azdeq.gov/wqd/reclaimed_water_rules_dec17.pdf.

———. "Draft Fact Sheet: Water Innovation Challenge—Pure Water Demonstration Facility. Individual Reclaimed Water Permit No. R512402." August 7, 2017. https://static.azdeq.gov/pn/fs_waterinnv.pdf.

———. "Draft Permit: State of Arizona Individual Reclaimed Water Permit No. R-512402." April 28, 2017. https://static.azdeq.gov/pn/draft_permit_water_innovation.pdf.

Arthur D. Little, Inc. "Flavor Profile Describes Food Flavors in Easily Understandable Terms." *Food Processing* 11, no. 5 (1950): 30–31.

———. "Food and Agribusiness Memorandum No. 2, 'Smell Fellows.'" n.d. Series VI, box 11, folder 5. MIT Archives, Cambridge, MA.

———. "Food and Agribusiness Memorandum No. 7, 'Measuring Odors.'" n.d. Series VI, Box 11, Folder 5. MIT Archives, Cambridge MA.

———. "Report on Flavor Studies of IPANA to Bristol-Myers Company." March 21, 1951. Series IV, box 45, Bound Technical Reports. MIT Archives, Cambridge MA.

Associated Press in Portland. "Sewer Brewers: Oregon Beer-Makers Challenged to Use Wastewater in Recipes." *Guardian*, April 29, 2015. www.theguardian.com/us-news/2015/apr/29/oregon-portland-beer-brew-sewer-water-competition.

Baehler, Karen J., Marquise McGraw, Michele J. Aquino, Ryan Heslin, Lindsay McCormick, and Tom Neltner. "Full Lead Service Line Replacement: A Case Study of Equity in Environmental Remediation." *Sustainability* 14, no. 1 (January 2022): 352. https://doi.org/10.3390/su14010352.

Bahng, Aimee. "The Pacific Proving Grounds and the Proliferation of Settler Environmentalism." *Journal of Transnational American Studies* 11, no. 2 (2020). https://doi.org/10.5070/T8112049580.

Baird, Davis. "Analytical Chemistry and the 'Big' Scientific Instrumentation Revolution." *Annals of Science* 50, no. 3 (1993): 267–90.

Bakker, Karen J. "A Political Ecology of Water Privatization." *Studies in Political Economy* 70, no. 1 (March 1, 2003): 35–58. https://doi.org/10.1080/07078552.2003.11827129.

Ballestero, Andrea. *A Future History of Water.* Durham, NC: Duke University Press, 2019.

Banister, Jeffrey M., and Stacie G. Widdifield. "The Debut of 'Modern Water' in Early 20th Century Mexico City: Xochimilco Potable Waterworks." *Journal of Historical Geography* 46 (2014): 36–52.

Barbieri, Maurizio, Marino Domenico Barberio, Francesca Banzato, Andrea Billi, Tiziano Boschetti, Stefania Franchini, Francesca Gori, and Marco Petitta. "Climate Change and Its Effect on Groundwater Quality." *Environmental Geochemistry and Health* 45 (2023): 1133–44. https://doi.org/10.1007/s10653-021-01140-5.

Barham, Elizabeth. "Translating Terroir: The Global Challenge of French AOC Labeling." *Journal of Rural Studies* 19, no. 1 (January 2003): 127–38. https://doi.org/10.1016/S0743-0167(02)00052-9.

Bartle, Keith, and Peter Myers. "History of Gas Chromatography." *Trends in Analytical Chemistry* 21, nos. 9–10 (2002): 547–57.

Barwich, Ann-Sophie. *Smellosophy: What the Nose Tells the Mind.* Cambridge, MA: Harvard University Press, 2020.

Bauby, Pierre. "Water Globalization: The Strategies of the Two French 'Majors.'" In *Globalized Water: A Question of Governance*, edited by Graciela Schneier-Madanes, 45–61. Dordrecht: Springer, 2014. https://doi.org/10.1007/978-94-007-7323-3_4.

Baylis, J. R. "Percolation and Runoff." *Journal (American Water Works Association)* 55 (1963): 40, 42.

———. "Preliminary Experiments on the Treatment of Lake Michigan Water for Chicago." *Journal (American Water Works Association)* 17 (1927): 710.

Baylis, John R., and C. L. Ehrhart. "Taste and Odor Elimination." *Journal (American Water Works Association)* 24, no. 5 (May 1932): 635–59.

Baylis, John R., Paul Hansen, R. A. Hoot, Robert Spurr Weston, Charles P. Hoover, W. A. Helbig, and E. B. Showell. "Further Observations on the Use of Activated Carbon in Removing Objectionable Taste and Odors from Water [with Discussion]." *Journal (American Water Works Association)* 22, no. 11 (November 1930): 1438–75.

Bebb, Herbert. "Voice of the People: Clean Water." *Chicago Daily Tribune (1923–1963)*, April 23, 1937. ProQuest Historical Newspapers.

Benson, Etienne. "Generating Infrastructural Invisibility: Insulation, Interconnection, and Avian Excrement in the Southern California Power Grid." *Environmental Humanities* 6 (2015): 103–30.

Benson, Etienne S. *Surroundings: A History of Environments and Environmentalisms.* Chicago: University of Chicago Press, 2020.

Bentley, Amy. *Inventing Baby Food: Taste, Health, and the Industrialization of the American Diet*. Berkeley: University of California Press, 2014.

Bentz, Julia. "Learning about Climate Change in, with and through Art." *Climatic Change* 162, no. 3 (October 1, 2020): 1595–1612. https://doi.org/10.1007/s10584-020-02804-4.

Berenstein, Nadia. "Designing Flavors for Mass Consumption." *Senses and Society* 13, no. 1 (2018): 19–40.

———. "Flavor Added." PhD diss., University of Pennsylvania, 2017.

Besky, Sarah. *The Darjeeling Distinction*. California Studies in Food and Culture. Berkeley: University of California Press, 2013.

———. "The Labor of Terroir and the Terroir of Labor: Geographical Indication and Darjeeling Tea Plantations." *Agriculture and Human Values* 31, no. 1 (March 1, 2014): 83–96. https://doi.org/10.1007/s10460-013-9452-8.

———. *Tasting Qualities: The Past and Future of Tea*. Berkeley: University of California Press, 2020.

Bess, Michael D. "Ecology and Artifice: Shifting Perceptions of Nature and High Technology in Postwar France." *Technology and Culture* 36, no. 4 (1995): 830–62. https://doi.org/10.2307/3106917.

Beutler, Lisa. "Getting the Public on Board with Your Potable Water Reuse Plan." *Water Resources IMPACT* 18, no. 4 (2016): 18–20.

Biénabe, Estelle, and Delphine Marie-Vivien. "Institutionalizing Geographical Indications in Southern Countries: Lessons Learned from Basmati and Rooibos." *World Development* 98 (October 1, 2017): 58–67. https://doi.org/10.1016/j.worlddev.2015.04.004.

Billen, C. F. Gilles, Josette Garnier, André Ficht, and Christine Cun. "Modeling the Response of Water Quality in the Seine River Estuary to Human Activity in Its Watershed over the Last 50 Years." *Estuaries*, no. 24 (2001): 977–93.

Billings, L. C. "Taste and Odor Control at Dallas." *Journal (American Water Works Association)* 32, no. 4 (1940): 613–20.

Biltekoff, Charlotte. *Eating Right in America: The Cultural Politics of Food and Health*. Durham, NC: Duke University Press, 2013.

Biltekoff, Charlotte, and Julie Guthman. "Conscious, Complacent, Fearful: Agri-Food Tech's Market-Making Public Imaginaries." *Science as Culture* 32, no. 1 (2023):58–82. https://doi.org/10.1080/09505431.2022.2090914.

Boelens, Rutgerd, Jaime Hoogesteger, Erik Swyngedouw, Jeroen Vos, and Philippus Wester. "Hydrosocial Territories: A Political Ecology Perspective." *Water International* 41, no. 1 (2016): 1–14. https://doi.org/10.1080/02508060.2016.1134898.

Bouhel, Zeineb, Jimmy Köpke, Mariam Mina, and Vladimir Smaktin. *Global Bottled Water Industry: A Review of Impacts and Trends.* Hamilton, ONT: United Nations University Institute for Water, Environment, and Health. 2023. https://inweh.unu.edu/global-bottled-water-industry-a-review-of-impacts-and-trends/.

Bouk, David. *How Our Days Became Numbered: Risk and the Rise of the Statistical Individual.* Chicago: University of Chicago Press, 2015.

Bourdieu, Pierre. *Distinction: A Social Critique of the Judgment of Taste.* Translated by Richard Nice. Cambridge, MA: Harvard University Press, 1987.

Bowen, Sarah. "Development from Within? The Potential for Geographical Indications in the Global South." *Journal of World Intellectual Property* 13, no. 2 (2010): 231–52. https://doi.org/10.1111/j.1747-1796.2009.00361.x.

———. *Divided Spirits: Tequila, Mezcal, and the Politics of Production.* Oakland: University of California Press, 2015.

Braidech, Matthew M., Lloyd C. Billings, F. Wellington Gilcreas, Neil Kershaw, R. D. Scott, and George R. Spaulding. "Specifications and Tests for Powdered Activated Carbon: Final Report of the Sub-committee." *Journal (American Water Works Association)* 30, no. 7 (1938): 1133–1224.

Braun, Bruce. "Biopolitics and the Molecularization of Life." *Cultural Geographies* 14 (2007): 6–28.

Bruchet, A., K. N'Guyen, J. Mallevialle, and C. Anselme. "Identification and Behaviour of Iodinated Haloform Medicinal Odor." In *Organic Micropollutants in the Aquatic Environment*, edited by G. Angeletti and A. Bjørseth, 371–83. Dordrecht: Springer Netherlands, 1991. https://doi.org/10.1007/978-94-011-3356-2_41.

Budahegyi, M. V., E. R. Lombosi, T. S. Lombosi, S. Y. Meszáros, Sz Nyiredy, G. Tarján, I Timár, and J. M. Takács. "Twenty-Fifth Anniversary of the Retention Index System in Gas-Liquid Chromatography." *Journal of Chromatography* 271 (1983): 213–307.

Bugger, Brad. "The INL and the Snake River Plain Aquifer." Idaho Department of Energy, Press Release, September 23, 2010. www.id.energy.gov/news/PressReleases/PR100923.htm.

Buller, Henry. "Privatization and Europeanization: The Changing Context of Water Supply in Britain and France." *Journal of Environmental Planning and Management* 39, no. 4 (1996): 461–82. https://doi.org/10.1080/09640569612336.

Burdick, Charles B. "Water Works Intakes." *Journal (American Water Works Association)* 38, no. 3 (1936): 315–25. www.jstor.org/stable/23347670.

Burgess, Ernest Watson, and Charles Shelton Newcomb, eds. *Census Data of the City of Chicago, 1930*. Chicago: University of Chicago Press, 1933.

Burlingame, Gary A., Roger M. Dann, and Geoffrey L. Brock. "A Case Study of Geosmin in Philadelphia's Water." *Journal (American Water Works Association)* 78, no. 3 (1986): 56–61.

Burlingame, Gary A., Djanette Khiari, and I. H. Suffet. "Odor Reference Standards: The Universal Language." *Proceedings: Water Quality Technology Conference* (1991): 7–25.

Butler, Ella. "Tasting Off-Flavors: Food Science, Sensory Knowledge and the Consumer Sensorium." *Senses and Society* 13, no. 1 (2018): 75–88.

Cain, Louis. "Raising and Watering a City: Ellis Sylvester Chesbrough and Chicago's First Sanitation System." *Technology and Culture* 13, no. 3 (1972): 353–72.

———. *Sanitation Strategy for a Lakefront Metropolis: The Case of Chicago*. DeKalb: Northern Illinois University Press, 1978.

Cairncross, Stanley E., and Loren B. Sjöström. "Flavor Profiles—A New Approach to Flavor Problems." *Food Technology* 4 (1950): 308–11.

Cantor, Alida. "Hydrosocial Hinterlands: An Urban Political Ecology of Southern California's Hydrosocial Territory." *Environment and Planning E: Nature and Space* 4, no. 2 (2021): 451–74. https://doi.org/10.1177/2514848620909384.

Carmichael, Wayne W., and Gregory L. Boyer. "Health Impacts from Cyanobacteria Harmful Algae Blooms: Implications for the North American Great Lakes." *Harmful Algae* 54 (April 2016): 194–212. https://doi.org/10.1016/j.hal.2016.02.002.

Carollo Engineers, Inc. *Long-Term Water Augmentation Options for Arizona: Prepared for the Long-Term Water Augmentation Committee of the Governor's Water Augmentation, Innovation, and Conservation Council*. August 2019. https://new.azwater.gov/sites/default/files/Long-Term%20Water%20Augmentation%20Options%20final.pdf.

Carruth, Allison, and Robert P. Marzec. "Environmental Visualization in the Anthropocene: Technologies, Aesthetics, Ethics." *Public Culture* 26, no. 2 (2014): 205–11.

Cartron, Bernard. "De la sècheresse." *Quotidien*, June 18, 1976. Série 5/7, boîte 11. Bibliothèque Historique de la Ville de Paris (BHVP).

Caul, Jean. "The Profile Method of Flavor Analysis." In *Advances in Food Research*, edited by E. M. Mrak and G. F. Stewart, 7:1–40. New York: Academic Press, 1957.

Cazaux, Maurice. "Dégustateurs en blanc pour une eau pure." *Le Figaro*, December 26, 1978. Série 5/7, boîte 11. Bibliothèque Historique de la Ville de Paris (BHVP).

Centrone, Ian. "20 Minutes With: Celebrated Water Sommelier Jessica Altieri." Penta. Accessed March 20, 2023. www.barrons.com/articles/20-minutes-with-celebrated-water-sommelier-jessica-altieri-01555344764.

Chang, Hasok. *Is Water H2O? Evidence, Realism and Pluralism*. New York: Springer, 2012.

Charles, Nicole. *Suspicion: Vaccines, Hesitancy, and the Affective Politics of Protection in Barbados*. Durham, NC: Duke University Press, 2022.

*Chicago Daily Tribune (1923–1963)*. "Filtration Plants." April 11, 1936. ProQuest Historical Newspapers.

———. "Makes City's Water Taste Like a Spring." April 17, 1930. ProQuest Historical Newspapers.

———. "So. Side Groups Oppose Water Filter Project: U.S. Loan Sought to Build Plant." January 13, 1934. ProQuest Historical Newspapers.

———. "South Side Fights Pollution Peril: Asks Increase of Flow into Calumet River." July 26, 1942. ProQuest Historical Newspapers.

———. "Sprague Urges Water Filtering to Remedy Taste: Industrial Phenol Pollution Held Accident." January 5, 1933. ProQuest Historical Newspapers.

———. "U.S. Keeps Eye on Thompson's Meter Proposal." April 18, 1927. ProQuest Historical Newspapers.

———. "Water Meters," August 26, 1927. ProQuest Historical Newspapers.

"Chicagwa Water." Quality Meats Creative. May 3, 2022. www.drinkchicagwa.com.

Chiu, Victoria. "Water Law in France." In *Encyclopedia of the Environment*. Article published February 11, 2019. www.encyclopedie-environnement.org/en/society/water-law-in-france/.

City of Scottsdale. "Aquifer Storage and Recovery Wells." 2022. www.google.com/maps/d/viewer?mid=1BOvmYOvda9vcW96LV5ULNLH19CEcgchP.

Classen, Constance, David Howes, and Anthony Synott. *Aroma: The Cultural History of Smell.* London: Routledge, 1994.

Claus, C. Anne. "Beyond Merroir: The Okinawan Taste for Clams." *Gastronomica* 17, no. 3 (2017): 49–57. https://doi.org/10.1525/gfc.2017.17.3.49.

Cohen, Benjamin R. *Pure Adulteration: Cheating on Nature in the Age of Manufactured Food.* Chicago: University of Chicago Press, 2020.

Colton, Craig E. "Industrial Wastes in Southeast Chicago: Production and Disposal 1870–1970." *Environmental Review: ER* 10, no. 2 (1986): 93–105. www.jstor.org/stable/3984560.

Conway, Brian D. "Land Subsidence and Earth Fissures in South-Central and Southern Arizona, USA." *Hydrogeology Journal* 24, no. 3 (May 1, 2016): 649–55. https://doi.org/10.1007/s10040-015-1329-z.

Corasaniti, Nick, Corey Kilgannon, and John Schwartz. "Tainted Water, Ignored Warnings and a Boss With a Criminal Past." *New York Times,* August 24, 2019, sec. New York. www.nytimes.com/2019/08/24/nyregion/newark-lead-water-crisis.html.

Corbin, Alain. *The Foul and Fragrant: Odor and the French Social Imagination.* Cambridge, MA: Harvard University Press, 1986.

Corsi, Steven R., David J. Graczyk, Steven W. Geis, Nathaniel L. Booth, and Kevin D. Richards. "A Fresh Look at Road Salt: Aquatic Toxicity and Water-Quality Impacts on Local, Regional, and National Scales." *Environmental Science & Technology* 44, no. 19 (2010): 7376–82. https://doi.org/10.1021/es101333u.

"Council Directive 98/83/EC of 3 November 1998 on the Quality of Water Intended for Human Consumption." November 3, 1998. Official Journal L 330, 05/12/1998, p. 0032-0054. http://data.europa.eu/eli/dir/1998/83/oj.

Crewe, Katherine. "Arizona Native Plants and the Urban Challenge." *Landscape Journal* 32, no. 2 (2013): 215–29.

Croijmans, Ilja, and Asifa Majid. "Not All Flavor Expertise Is Equal: The Language of Wine and Coffee Experts." *PLOS ONE* 11, no. 6 (June 20, 2016): e0155845. https://doi.org/10.1371/journal.pone.0155845.

Croissant, Jennifer. "Agnotology: Ignorance and Absence or Towards a Sociology of Things That Aren't There." *Social Epistemology* 28, no. 1 (2014): 4–25.

Cronon, William. *Nature's Metropolis: Chicago and the Great West.* New York: W. W. Norton, 1991.

Daston, Lorraine J., and Peter Galison. *Objectivity.* Cambridge, MA: MIT Press, 2010.

Daugherty, E. N., A. V. Ontiveros-Valencia, J. S. Rice, M. J. Wiest, and R. U. Halden. "Impact of Point-of-Use Water Softening on Sustainable Water Reclamation: Case Study of the Greater Phoenix Area." In *Contaminants of Emerging Concern in the Environment: Ecological and Human Health Considerations,* 1048:497–518. ACS Symposium Series 1048. American Chemical Society, 2010. https://doi.org/10.1021/bk-2010-1048.ch025.

Davis, Katrinell M. *Tainted Tap: Flint's Journey from Crisis to Recovery.* Chapel Hill: University of North Carolina Press, 2021.

Dawson, Kevin. *Undercurrents of Power: Aquatic Culture in the African Diaspora.* Philadelphia: University of Pennsylvania Press, 2018.

De Vries, Hanna, Victor Bekkers, and Lars Tummers. "Innovation in the Public Sector: A Systematic Review and Future Research Agenda." *Public Administration* 94, no. 1 (2016): 146–66. https://doi.org/10.1111/padm.12209.

DeJong, David H. "A Scheme to Rob Them of Their Land: Water, Allotment, and the Economic Integration of the Pima Reservation, 1902–1921." *Journal of Arizona History* 44, no. 2 (2003): 99–132.

Dennis, Carolyn. "Judges Pick Olympia, Wash., Winner in Eight-City Taste Testing." June 23, 1985. Associated Press News Archive.

Department of Public Works, Bureau of Engineering, Division of Water Purification. *How Chicago Protects the Quality of Its Water Supply.* Chicago: Division of Water Purification, 1940.

DeSalle, Rob, and Ian Tattersalle. *A Natural History of Beer.* New Haven, CT: Yale University Press, 2019.

Dettinger, Michael, Bradley Udall, and Aris Georgakakos. "Western Water and Climate Change." *Ecological Applications* 25, no. 8 (2015): 2069–93. https://doi.org/10.1890/15-0938.1.

Doherty, James. "South Siders Owe a Lot to This Stranger: Thanks to Him, Their Water Is Pure." *Chicago Daily Tribune (1923–1963),* May 1, 1949, 30. ProQuest Historical Newspapers.

———. "Thompson Defies War Officials in Meter Fight: Calls Ordinance Crooked and Inspired by Graft." *Chicago Daily Tribune (1923–1963),* September 12, 1927, 19. ProQuest Historical Newspapers.

Donaldson, Wellington. "Chlorination Tastes and Odors." *Journal (American Water Works Association)* 9, no. 6 (November 1922): 885–91.

Douglas, Mary. *Purity and Danger.* New York: Routledge, 1966.

Dow, Cory, Sajjad Ahmad, Krystyna Stave, and Daniel Gerrity. "Evaluating the Sustainability of Indirect Potable Reuse and Direct Potable Reuse: A Southern Nevada Case Study." *AWWA Water Science* 1, no. 4 (2019): e1153. https://doi.org/10.1002/aws2.1153.

Dravnieks, Andrew. "Odor Quality: Semantically Generated Multidimensional Profiles Are Stable." *Science* 218, no. 4574 (1982): 799–801. https://doi.org/10.1126/science.7134974.

Dravnieks, Andrew, and Anne O'Donnell. "Principles and Some Techniques of High-Resolution Headspace Analysis." *Journal of Agricultural and Food Chemistry* 19, no. 6 (1971): 1049–56.

Dunne, Anthony, and Fiona Raby. *Speculative Everything: Design, Fiction, and Social Dreaming.* Cambridge, MA: MIT Press, 2013.

DuPuis, E. Melanie. *Nature's Perfect Food: How Milk Became America's Drink.* New York: New York University Press, 2002.

DuPuis, François. "Equipement: L'Eau à vau-l'eau." *L'Express* 20–26 (July 1970): 56–57. Série 5/7, boîte 11. Bibliothèque Historique de la Ville de Paris (BHVP).

E. B., J. "Filtered Water." *Chicago Daily Tribune (1923–1963)*, February 10, 1935. ProQuest Historical Newspapers.

Earl, G. G. "Minutes." *Proceedings, Thirty-Fifth Annual Convention American Water Works Association [with Discussion]* 2 (1915): 251.

Ellms, J. W., and W. C. Lawrence. "The Causes of Obnoxious Tastes and Odors Sometimes Occurring in the Cleveland Water Supply." *Journal (American Water Works Association)* 9 (1922): 463–73.

Elsaid, Khaled, Mohammed Kamil, Enas Taha Sayed, Mohammad Ali Abdelkareem, Tabbi Wilberforce, and A. Olabi. "Environmental Impact of Desalination Technologies: A Review." *Science of the Total Environment* 748 (December 15, 2020): 141528. https://doi.org/10.1016/j.scitotenv.2020.141528.

Environmental Protection Agency. Announcement of Preliminary Regulatory Determinations for Contaminants on the Fourth Drinking Water Contaminant Candidate List. 86 Fed. Reg. 12272,, 40 C.F.R. 141 (2020). www.federalregister.gov/documents/2020/03/10/2020-04145/announcement-of

-preliminary-regulatory-determinations-for-contaminants-on-the-fourth -drinking-water.

Ericson, John. *The Quality Problem in Relation to Chicago's Water Supply: Official Report to Col. A. A. Sprague, Commissioner of Public Works*. Chicago, IL, May 1925.

———. "Universal Metering of Water Supplies." *Journal (American Water Works Association)* 16, no. 4 (1926): 399–414.

———. *The Water Supply System of Chicago*. Chicago: Barnard & Miller Print, April 1924.

Espeland, Wendy Nelson. *The Struggle for Water: Politics, Rationality, and Identity in the American West*. Chicago: University of Chicago Press, 1998.

Ettre, L. A. "The Kovats Retention Index System." *Analytical Chemistry* 36, no. 8 (July 1964): 31A–41A.

European Chemicals Agency. "Perfluoroalkyl Chemicals (PFASs)—ECHA." Accessed December 19, 2022. https://echa.europa.eu/en-US/hot-topics/perfluoroalkyl-chemicals-pfas.

European Drinking Water Secretariat. "European Drinking Water: History." Accessed July 3, 2023. www.europeandrinkingwater.eu/initiative/background/history/.

Euzen, Agathe. "Voir, goûter, sentir . . . perceptions de la qualité de l'eau par les Parisiens." In *L'Eau mondialisée: la Gouvernance en question*, edited by Graciela Schneier-Madenes, 471–83. Paris: La Découverte, 2010.

Evans, Arthur. "Test Shows How Waste of Water Cuts Pressure: Metering Is Only Solution to Problem." *Chicago Daily Tribune (1923–1963)*, February 14, 1925. ProQuest Historical Newspapers.

Evans, W. A. "How to Keep Well: Can Water Be Made Safe?" *Chicago Daily Tribune (1923–1963)*, July 11, 1933. ProQuest Historical Newspapers.

Fales, Almon L., Wellington Donaldson, C. A. Emerson, Harry B. Hommon, F. W. Mohlman, James A. Newlands, and Ralph Hilscher. "Progress Report of Committee on Industrial Wastes in Relation to Water Supply." *Journal (American Water Works Association)* 10, no. 3 (1923): 415–30.

Ferguson, Dean T. "Nightsoil and the 'Great Divergence': Human Waste, Urban Economy, and Economic Productivity, 1500–1900." *Journal of Global History* 9 (2014): 379–402.

Filby, E. L. "Water Supplies of Florida." *Journal (American Water Works Association)* 17, no. 1 (1927): 37–50.

"Fine Water Academy." Accessed December 16, 2022. http://finewateracademy.com/.

Fitch, Kimberly A. "Liquidating the Public Sector? Water Privatization in France and Germany." PhD diss., University of Pennsylvania, 2007. https://search.proquest.com/openview/9d940a77e50475d8294820Idf99db1a3/1?pq-origsite=gscholar&cbl=18750&diss=y.

Flanigan, Kathy. "Beer from MMSD Wastewater? Engineer's 'Sludge' Brew Has People Talking." *Milwaukee Journal Sentinel*, January 22, 2015. https://archive.jsonline.com/entertainment/beer/engineer-makes-beer-with-mmsd-water-b99427598z1-289463111.html/.

Flentje, Martin E., and H. H. Gerstein. "Report of Committee on Control of Tastes and Odors [with Discussion]." *Journal (American Water Works Association)* 24, no. 11 (November 1932): 1738–49.

Forrest, Sherrie, and Michael A. Feder. *Climate Change Education: Goals, Audiences, and Strategies*. Washington, DC: National Academies Press, 2011.

Fortun, Kim. "Ethnography in Late Industrialism." *Cultural Anthropology* 27 (2012): 446–64. https://doi.org/10.1111/j.1548-1360.2012.01153.x.

Fountain, Henry. "In a First, U.S. Declares Shortage on Colorado River, Forcing Water Cuts." *New York Times*, August 16, 2021, sec. Climate. www.nytimes.com/2021/08/16/climate/colorado-river-water-cuts.html.

Freeman, Scott. "Perfume and Planes: Ignorance and Imagination in Haiti's Vetiver Oil Industry." *Journal of Latin American and Caribbean Anthropology* 24, no. 1 (2019): 110–26. https://doi.org/10.1111/jlca.12368.

Freidberg, Susanne. *Fresh: A Perishable History*. Cambridge, MA: Harvard University Press, 2010.

Fretwell, Erica. *Sensory Experiments: Psychophysics, Race, and the Aesthetics of Feeling*. Durham, NC: Duke University Press, 2020.

Frickel, Scott, and M. Bess Vincent. "Hurricane Katrina, Contamination, and the Unintended Organization of Ignorance." In "Perspectives on Hurricane Katrina," edited by A. George Schillinger, special issue, *Technology in Society* 29, no. 2 (2007): 181–88. https://doi.org/10.1016/j.techsoc.2007.01.007.

Frolich, Xaq. "The Informational Turn in Food Politics: The US FDA's Nutrition Label as Information Infrastructure." *Social Studies of Science* 47, no. 2 (2017): 145–71. https://doi.org/10.1177/0306312716671223.

Fuller, George H., R. Steltenkamp, and G. A. Tisserand. "The Gas Chromatograph with Human Sensor: Perfumer Model." *Annals of the New York Academy of Sciences* 116 (1964): 711–24.

Gal, Ofer, and Raz Chen-Morris. "Empiricism without the Senses: How the Instrument Replaced the Eye." In *The Body as Object and Instrument of Knowledge: Embodied Empiricism in Early Modern Science*, edited by Charles T. Wolfe and Ofer Gal, 121–47. Studies in History and Philosophy of Science. Dordrecht: Springer Netherlands, 2010. https://doi.org/10.1007/978-90-481-3686-5_7.

Galison, Peter. "Removing Knowledge." *Critical Inquiry* 31, no. 1 (2004): 229–43.

Gamble, Eleanor Acheson McCulloch. "The Applicability of Weber's Law to Smell." *American Journal of Psychology* 10, no. 1 (October 1898): 82–142.

Gandy, Matthew. *The Fabric of Space: Water, Modernity, and the Urban Imagination*. Cambridge, MA: MIT Press, 2014.

———. "Rethinking Urban Metabolism: Water, Space, and the Modern City." *City* 8, no. 3 (2004): 363–79.

Garcier, Romain. "The Placing of Matter: Industrial Water Pollution and the Construction of Social Order in Nineteenth-Century France." *Journal of Historical Geography* 36 (2010): 132–42.

Gatrell, Jay D., David Nemeth, and Charles D. Yeager. "Sweetwater, Mountain Springs, and Great Lakes: A Hydro-Geography of Beer Brands." In *The Geography of Beer Regions, Environment, and Societies*, edited by Mark Patterson and Nancy Hoalst-Pullen, 89–98. New York: Springer, 2014.

Gayton, Loran D., and Arthur E. Gorman. "The World's Largest Filtration Plant [with Discussion]." *Journal (American Water Works Association)* 32, no. 8 (1940): 1359–80.

Gerber, N., and H. Lechevalier. "Geosmin, an Earthy-Smelling Substance Isolated from Actinomycetes." *Applied Microbiology* 13, no. 6 (November 1965): 935–38.

Gerkin, Richard C., Kathrin Ohla, Maria G. Veldhuizen, Paule V. Joseph, Christine E. Kelly, Alyssa J. Bakke, Kimberley E. Steele, et al. "Recent Smell Loss Is the Best Predictor of COVID-19 among Individuals with Recent Respiratory Symptoms." *Chemical Senses* 46 ( 2021). https://doi.org/10.1093/chemse/bjaa081.

"Germany: Waste Water Beer Is Defying Purity Law." Inside Beer. June 18, 2019. www.inside.beer/news/detail/germany-waste-water-beer-is-defying-purity-law.html.

Gerontas, Apostolos. "Creating New Technologists of Research in the 1960s: The Case of the Reproduction of Automated Chromatography Specialists and Practitioners." *Science and Education* 23 (2014): 1681–1700.

Geroski, Paul A. "Models of Technology Diffusion." *Research Policy* 29, nos. 4–5 (April 2000): 603–25. https://doi.org/10.1016/S0048-7333(99)00092-X.

Gerstein, H. H. "Chloro-Phenol Tastes and Abnormal Absorption of Chlorine in the Chicago Water Supply." *Journal (American Water Works Association)* 21, no. 3 (1929): 346–57.

Gerstein, Hyman H. "Lake Michigan Pollution and Chicago's Supply." *Journal (American Water Works Association)* 57, no. 7 (1965): 841–57.

Gibbons, Mortimer M. "Water Pollution by Petroleum Oils." *Journal (American Water Works Association)* 32, no. 3 (1940): 465–77.

Gill, Rosalind, and Andy Pratt. "In the Social Factory? Immaterial Labour, Precariousness and Cultural Work." *Theory, Culture & Society* 25, nos. 7–8 (2008): 1–30.

Gleich, Amy. "Wastewater Partnership Keeps Scottsdale's Golf Courses in the Green." *KTAR*, November 6, 2013. https://ktar.com/story/78879/wastewater-partnership-keeps-scottsdales-golf-courses-in-the-green/.

Gleick, Peter H. *Bottled and Sold: The Story Behind Our Obsession with Bottled Water.* Washington, DC: Island Press, 2010.

Goldbaum, Christina. "'Tasting Funny for Years': Lead in the Water and a City in Crisis." *New York Times*, August 20, 2019, sec. New York. www.nytimes.com/2019/08/20/nyregion/newark-water-crisis.html.

Gorman, Arthur E. "Chicago's South District Water Filtration Project. Historical Aspects—Development of the PWA Project—General Description of the Project as Proposed by the Water Purification Division." January 16, 1939. Chicago Public Library.

Goubert, Jean-Pierre. *La conquête de l'eau: L'avenement de la santé à l'âge industriel.* Paris: Editions Robert Laffont, 1986.

Greeley, Samuel A. "Operation of the Water Filtration Plants at Evanston, Illinois, and Whiting, Indiana." *Journal (American Water Works Association)* (July 1921): 442–47.

Grob, K. "Organic Substances in Water and in Its Precursor, Part I: Methods for Their Determination by Gas-Liquid Chromatography." *Journal of Chromatography A* 84, no. 2 (1973): 255–73.

Grob, K., K. Grob Jr., and G. Grob. "Organic Substances in Potable Water and in Its Precursor, III: The Closed-Loop Stripping Procedure Compared with Rapid Liquid Extraction." *Journal of Chromatography A* 106, no. 2 (1975): 299–315.

Grove, Andrew S. "Navigating Strategic Inflection Points." *Business Strategy Review* 19, no. 3 (September 2008): 57–63. https://doi.org/10.1111/1467-8616.00033.

Gullans, Oscar. "The Comparison of Odor Elimination Treatments." *Journal (American Water Works Association)* 29, no. 1 (1937): 60–65.

———. "Procedure for Making Odor Determinations." *Journal (American Water Works Association)* 25, no. 7 (July 1933): 974–81.

Guy, Kolleen. *When Champagne Became French: Wine and the Making of a National Identity*. Baltimore, MD: Johns Hopkins University Press, 2003.

Hager, Alex. "Arizona, Nevada and Mexico See Water Usage Cuts as Colorado River Shortage Deepens." *NPR*, August 16, 2022, sec. National. www.npr.org/2022/08/16/1117762260/arizona-nevada-and-mexico-see-water-usage-cuts-as-colorado-river-shortage-deepen.

Hamlin, Christopher. *A Science of Impurity: Water Analysis in Nineteenth Century Britain*. Berkeley: University of California Press, 1990.

———. "'Waters' or 'Water'?—Master Narratives in Water History and Their Implications for Contemporary Water Policy." *Water Policy* 2 (2000): 313–25.

Hansen, Paul, Charles H. Spaulding, Roberts Hulbert, Lewis I. Birdsall, George D. Norcom, and John R. Baylis. "Tastes and Odors in Public Water Supplies, Causes and Remedies [with Discussion]." *Journal (American Water Works Association)* 23, no. 10 (October 1931): 1495–1509.

Harbison, Martha. "Daily Infographic, Beer Edition: The Beer Flavor and Aroma Wheel." *Popular Science*, January 10, 2013. www.popsci.com/science/article/2013-01/infographic-day-beersci-edition-beer-flavor-wheel/.

Harlow, Ivan F., Thomas J. Powers, and Ralph B. Ehlers. "The Phenolic Waste Treatment Plant of the Dow Chemical Company." *Sewage Works Quarterly* 10, no. 6 (November 1938): 1048–59.

Harper, Roland. "Some Chemicals Representing Particular Odour Qualities." *Chemical Senses and Flavor* 1 (1975): 353–57.

Harrison, Louis B., and A. S. Behrman. "Use of Granular Activated Carbon at Bay City [with Discussion]." *Journal (American Water Works Association)* 32, no. 4 (1940): 593–608.

Hawkins, Gay. "Commentary—Making Water into a Political Material: The Case of PET Bottles." *Environment and Planning A* 43 (2011): 2001–6.

———. "The Politics of Bottled Water." *Journal of Cultural Economy* 2, nos. 1–2 (2009): 183–95.

Hayes-Conroy, Allison, and Deborah G. Martin. "Mobilising Bodies: Visceral Identification in the Slow Food Movement." *Transactions of the Institute of British Geographers* 35, no. 2 (2010): 269–81.

Hayes-Conroy, Jessica, and Allison Hayes-Conroy. "Visceral Geographies: Mattering, Relating, and Defying." *Geography Compass* 4, no. 9 (2010): 1273–83. https://doi.org/10.1111/j.1749-8198.2010.00373.x.

Heidelberger, Michael. *Nature from Within: Gustav Theodor Fechner and His Psychophysical Worldview.* Translated by Cynthia Klohr. Pittsburgh, PA: University of Pittsburgh Press, 2004.

Heinz, D. E., Rose Marie Pangborn, and W. G. Jennings. "Pear Aroma: Relation of Instrumental and Sensory Techniques." *Journal of Food Science* 29, no. 6 (1964): 756–61.

Helbig, W. A. "Comparison of Tests for Activated Carbons." *Journal (American Water Works Association)* 30, no. 7 (1938): 1225–33.

Heymann, Hildegarde. "A Personal History of Sensory Science." *Food, Culture, and Society* 22, no. 2 (2019): 203–23.

Hisano, Ai. *Visualizing Taste: How Business Changed the Look of What You Eat.* Cambridge, MA: Harvard University Press, 2019.

"History: Western Municipal Water District, CA." Accessed December 17, 2022. www.wmwd.com/88/History.

Hobart, Hi'ilei. "At Home on the Mauna: Ecological Violence and Fantasies of Terra Nullius on Maunakea's Summit." *Native American and Indigenous Studies* 6, no. 2 (2019): 30–50.

Hobart, Hi'ilei Julia Kawehipuaakahaopulani. *Cooling the Tropics: Ice, Indigeneity, and Hawaiian Refreshment.* Durham, NC: Duke University Press, 2022.

Hofste, Rutger Willem, Paul Reig, and Leah Schleifer. "17 Countries, Home to One-Quarter of the World's Population, Face Extremely High Water Stress." World Resources Institute, August 6, 2019. www.wri.org/insights/17-countries-home-one-quarter-worlds-population-face-extremely-high-water-stress.

Holness, D. "Gas-Liquid Chromatography and the Perfumer." *Journal of the Society of Cosmetic Chemists* 12, no. 7 (1961): 370–97.

Hommes, Lena, and Rutgerd Boelens. "Urbanizing Rural Waters: Rural-Urban Water Transfers and the Reconfiguration of Hydrosocial Territories in Lima." *Political Geography* 57 (March 1, 2017): 71–80. https://doi.org/10.1016/j.polgeo.2016.12.002.

Hooper, Jennifer, Denise Funk, Kati Bell, Morayo Noibi, Kyle Vickstrom, Chris Schulz, Eddie Machek, and Ching-Hua Huang. "Pilot Testing of Direct and Indirect Potable Water Reuse Using Multi-Stage Ozone-Biofiltration without Reverse Osmosis." *Water Research* 169 (February 1, 2020): 115178. https://doi.org/10.1016/j.watres.2019.115178.

Hoover, Elizabeth. *The River Is in Us: Fighting Toxics in a Mohawk Community.* Minneapolis: University of Minnesota Press, 2017.

Hough, Guillermo. *Sensory Shelf Life Estimation of Food Products.* Boca Raton, FL: CRC Press, 2010. https://doi.org/10.1201/9781420092943.

Howes, David, and Marc Lalonde. "The History of Sensibilities: Of the Standard of Taste in Mid-Eighteenth Century England and the Circulation of Smells in Post-Revolutionary France." *Dialectical Anthropology* 16, no. 2 (June 1, 1991): 125–35. https://doi.org/10.1007/BF00250241.

Howson, L. R. "Responsibility for Potability of Water Delivered." *Journal AWWA* 26, no. 9 (1934): 1146–51.

Hughes, Justin. "Coffee and Chocolate—Can We Help Developing Country Farmers through Geographical Indications?" SSRN Scholarly Paper. Rochester, NY, September 29, 2010. https://doi.org/10.2139/ssrn.1684370.

Hui, Alexandra. *The Psychophysical Ear: Musical Experiments, Experimental Sounds, 1840–1910.* Cambridge, MA: MIT Press, 2012.

Huisman, John L. A., and Asifa Majid. "Psycholinguistic Variables Matter in Odor Naming." *Memory & Cognition* 46, no. 4 (May 1, 2018): 577–88. https://doi.org/10.3758/s13421-017-0785-1.

Hyslop, J. H. "Psychophysics." *Science* 8, no. 189 (1886): 259–62.

Ibáñez Martín, Rebeca, and Fenna Smits. "The Material Politics of Living in Close Proximity to Our Wastewaters: A Case of Decentralisation in the Netherlands." *Environment and Planning E: Nature and Space*, October 17, 2022. https://doi.org/10.1177/25148486221131191.

Igo, Sarah E. *The Averaged American: Surveys, Citizens, and the Making of a Mass Public*. Cambridge, MA: Harvard University Press, 2007.

Illich, Ivan. *H2O and the Waters of Forgetfulness: Reflections on the Historicity of "Stuff"*. Dallas, TX: Dallas Institute of Humanities and Culture, 1985.

Institut National de l'Audiovisuel. "René Dumont: 'Je bois devant vous un verre d'eau précieuse.'" Video. Official electoral campaign, presidential election, first round, April 19, 1974. Accessed July 28, 2020. http://www.ina.fr/video/I09167743.

Izaguirre, George, Cordelia J. Hwang, Stuart W. Krasner, and Michael J. McGuire. "Geosmin and 2-Methylisoborneol from Cyanobacteria in Three Water Supply Systems." *Applied and Environmental Microbiology* 43, no. 3 (March 1, 1982): 708–14.

Jackson, Deborah Davis. "Scents of Place: The Dysplacement of a First Nations Community in Canada." *American Anthropologist* 113, no. 4 (2011): 606–18.

Jackson, Leland. "We Brewed Beer from Recycled Wastewater—and It Tasted Great." The Conversation, November 16, 2020. http://theconversation.com/we-brewed-beer-from-recycled-wastewater-and-it-tasted-great-148386.

Jas, Nathalie. "Public Health and Pesticide Regulation in France before and after Silent Spring." *History and Technology* 23, 4 (2007): 369–88.

Jennings, Eric T. *Curing the Colonizers: Hydrotherapy, Climatology, and French Colonial Spas*. Durham, NC: Duke University Press, 2006.

Jennings, Ralph. "Wastewater on a Stick: The Popsicles from Taiwan That You Really, Really Don't Want to Eat." *Los Angeles Times*, July 16, 2017, sec. World & Nation. www.latimes.com/world/asia/la-fg-taiwan-tainted-popsicles-2017-story.html.

Jennings, W. G., S. Leonard, and Rose Marie Pangborn. "Volatiles Contributing to the Flavor of Bartlett Pears." *Food Technology* 14 (1960): 587–90.

Jickling, Hannah, and Helen Reed. *Multiple Elementary*. London: Black Dog Publishing, 2018.

Jüttner, Friedrich, and Susan B. Watson. "Biochemical and Ecological Control of Geosmin and 2-Methylisoborneol in Source Waters." *Applied and Environmental Microbiology* 73, no. 14 (2007): 4395–4406. https://doi.org/10.1128/AEM.02250-06.

Kafka, Ben. "Ingestion/Power Hungry: Dining with the Committee of Public Safety." *Cabinet*, no. 32 (Winter 2008–2009). www.cabinetmagazine.org/issues/32/kafka.php.

Kartit, Dina. "Veolia's Nine-Month Revenue Jumps Nearly 50% on Suez Acquisition." Reuters, November 9, 2022. www.reuters.com/business/frances-veolias-nine-month-revenue-jumps-nearly-50-suez-acquisition-2022-11-09/.

Keating, Ann Durkin, Kathleen A. Brosnan, and William C. Barnett, eds. *City of Lake and Prairie: Chicago's Environmental History*. Pittsburgh, PA: University of Pittsburgh Press, 2020.

Kelley, Lindsay. "Tasting History: Biscuits, Culture, and National Identity." A Participatory Taste Workshop, November 21, 2019. https://cementa.com.au/.

Kendall, David A., and Anne J. Neilson. "Correlation of Subjective and Objective Odor Responses." *Annals of the New York Academy of Sciences* 116 (July 1964): 567–75.

Kenner, Alison, Aftab Mirzaei, and Christy Spackman. "Breathing in the Anthropocene: Thinking through Scale with Containment Technologies." *Cultural Studies Review* 25, no. 2 (December 2019): 153.

Kettler, Andrew. *The Smell of Slavery: Olfactory Racism and the Atlantic World*. Cambridge: Cambridge University Press, 2020.

Khaleefa, Omar. "Who Is the Founder of Psychophysics and Experimental Psychology?" *American Journal of Islamic Social Science* 16, no. 2 (1999): 1–26. https://doi.org/10.35632/ajis.v16i2.2126.

Khandelwal, Meena, Matthew E. Hill, Paul Greenough, Jerry Anthony, Misha Quill, Marc Linderman, and H. S. Udaykumar. "Why Have Improved Cook-Stove Initiatives in India Failed?" *World Development* 92 (April 1, 2017): 13–27. https://doi.org/10.1016/j.worlddev.2016.11.006.

Khiari, Djanette. *Distribution Generated Taste-and-Odor Phenomena*. Denver, CO: AWWA Research Foundation, 2002.

Kiechle, Melanie A. *Smell Detectives: An Olfactory History of Nineteenth-Century Urban America*. Weyerhauser Environmental Books. Seattle: University of Washington Press, 2017.

Kinchy, Abby, Sarah Parks, and Kirk Jalbert. "Fractured Knowledge: Mapping the Gaps in Public and Private Water Monitoring Efforts in Areas Affected by Shale Gas Development." *Environment and Planning C* 34 (2016): 879–99.

King, Alexandra. "The Aesthetic Attitude." In *The Internet Encyclopedia of Philosophy*. Accessed February 10, 2015. https://iep.utm.edu/aesth-at/.

Knorr Cetina, Karin. *Epistemic Cultures: How the Sciences Make Knowledge*. Cambridge, MA: Harvard University Press, 1999.

Konig, Michael. "Phoenix in the 1950s Urban Growth in the 'Sunbelt.'" *Arizona and the West* 24, no. 1 (1982): 19–38.

Kopp, Peter A. *Hoptopia: A World of Agriculture and Beer in Oregon's Willamette Valley*. Oakland: University of California Press, 2016.

Korsmeyer, Carolyn. *Aesthetics: The Big Questions*. Malden, MA: Wiley-Blackwell, 1998.

———. *Making Sense of Taste: Food and Philosophy*. Ithaca, NY: Cornell University Press, 1999.

———. *Savoring Disgust: The Foul and the Fair in Aesthetics*. New York: Oxford University Press, 2011.

Krasner, Stuart, Michael McGuire, and Virginia Ferguson. "Tastes and Odors: The Flavor Profile Method." *Journal (American Water Works Association)* 77, no. 3 (March 1985): 34–39.

Kristeva, Julia. *Powers of Horror: An Essay on Abjection*. Translated by Leon S. Roudiez. New York: Columbia University Press, 1982.

Kuhn, Eric, and John Fleck. *Science Be Dammed: How Ignoring Inconvenient Science Drained the Colorado River*. Tucson: University of Arizona Press, 2019.

"La Bataille marketing des eaux en bouteille et de l'eau du robinet." November 27, 2012. www.sircome.fr/la-bataille-marketing-des-eaux-en.

Lagard, Dorothée, Fabienne Waks, and Sandrine Alouf. *Realms of Water: A Journey of Discovery to the Sources of Mineral Water*. Paris: Editions Textuel, 2010.

Lahne, Jacob. "Sensory Science, the Food Industry, and the Objectification of Taste." *Anthropology of Food*, no. 10 (2016). http://aof.revues.org/7956.

———. "Standard Sensations: The Production of Objective Experience from Industrial Technique." *Senses and Society* 13 (2018). https://doi.org/10.1080/17458927.2017.1420842.

Lahne, Jacob, and Christy Spackman, eds. "Accounting for Taste: Technologies for Capturing Food-Sensory Experience." *Senses and Society* 13, no. 1 (March 2018): 1–5. https://doi.org/10.1080/17458927.2018.1427361.

Landecker, Hannah. "Food as Exposure: Nutritional Epigenetics and the New Metabolism." *BioSocieties* 6, no. 2 (June 2011): 167–94.

———. "The Social as Signal in the Body of Chromatin." *Sociological Review Monographs* 64, no. 1 (2016): 79–99.

Langergraber, Günter, and Elke Muellegger. "Ecological Sanitation—a Way to Solve Global Sanitation Problems?" *Environment International* 31, no. 3 (April 1, 2005): 433–44. https://doi.org/10.1016/j.envint.2004.08.006.

Langstaff, Susan A., and M. J. Lewis. "The Mouthfeel of Beer—a Review." *Journal of the Institute of Brewing* 99, no. 1 (1993): 31–37. https://doi.org/10.1002/j.2050-0416.1993.tb01143.x.

Larkin, Brian. "The Politics and Poetics of Infrastructure." *Annual Review of Anthropology* 42 (2013): 327–43.

Latour, Bruno. *Pandora's Hope: Essays on the Reality of Science Studies*. Cambridge, MA: Harvard University Press, 1999.

———. *The Pasteurization of France*. Cambridge, MA: Harvard University Press, 1993.

———. "Visualization and Cognition: Thinking with Eyes and Hands." *Knowledge and Society: Studies in the Sociology of Culture Past and Present* 6 (1986): 1–40.

Lawless, Harry T., and Hildegarde Heymann. "Measurement of Sensory Thresholds." In *Sensory Evaluation of Food: Principles and Practices*, edited by Harry T. Lawless and Hildegarde Heymann, 173–207. Boston, MA: Springer US, 1999. https://doi.org/10.1007/978-1-4615-7843-7_6.

———. *Sensory Evaluation of Food: Principles and Practices*. 2nd ed. New York: Springer, 2010.

Lazzarto, Mauricio. "Immaterial Labour." In *Radical Thought in Italy: A Potential Politics*, 133–47. Minneapolis: University of Minnesota Press, 1996.

Lecocq, Sébastien, and Michael Visser. "What Determines Wine Prices: Objective vs. Sensory Characteristics." *Journal of Wine Economics* 1, no. 1 (2006): 42–56. https://doi.org/10.1017/S1931436100000080.

Lemasson, Jean-Pierre, and Amy Trubek. "Terroir Products in North America: Dreams or Future Reality?" *Cuizine : Journal of Canadian Food Cultures/Cuizine : Revue des cultures culinaires au Canada* 2, no. 2 (2010). https://doi.org/10.7202/044347ar.

Lemov, Rebecca. "'Hypothetical Machines': The Science Fiction Dreams of Cold War Social Science." *Isis* 101, no. 2 (2010): 401–11. https://doi.org/10.1086/653107.

Leysne, Wendy L. H. "Journeys through 'Ingestible Topography'; Socializing the 'Situated Eater' in France." *European Studies* 22 (2006): 129–58.

Liboiron, Max. *Founder/Worker.* 2011. Trash, mixed media. https://maxliboiron.com/2013/08/06/founderworker/.

———. *Pollution Is Colonialism*. Durham, NC: Duke University Press, 2021.

———. "Redefining Pollution: Plastics in the Wild." PhD diss., New York University, 2012.

Liboiron, Max, and Josh Lepawsky. *Discard Studies: Wasting, Systems, and Power.* Cambridge, MA: MIT Press, 2022.

Liboiron, Max, Alex Zahara, Kaitlyn Hawkins, Christina Crespo, Bárbara de Moura Neves, Vonda Wareham-Hayes, Evan Edinger, et al. "Abundance and Types of Plastic Pollution in Surface Waters in the Eastern Arctic (Inuit Nunangat) and the Case for Reconciliation Science." *Science of the Total Environment* 782 (August 15, 2021): 146809. https://doi.org/10.1016/j.scitotenv.2021.146809.

Lin, Qi, Ke Zhang, Suzanne McGowan, Eric Capo, and Ji Shen. "Synergistic Impacts of Nutrient Enrichment and Climate Change on Long-Term Water Quality and Ecological Dynamics in Contrasting Shallow-Lake Zones." *Limnology and Oceanography* 66, no. 9 (2021): 3271–86. https://doi.org/10.1002/lno.11878.

Linton, Jamie. *What Is Water? The History of a Modern Abstraction*. Vancouver: University of British Columbia Press, 2010.

Lobo-Prieto, Ana, Noelia Tena, Ramón Aparicio-Ruiz, María T. Morales, and Diego L. García-González. "Tracking Sensory Characteristics of Virgin Olive Oils during Storage: Interpretation of Their Changes from a Multiparametric Perspective." *Molecules* 25, no. 7 (2020): 1686. https://doi.org/10.3390/molecules25071686.

Lorrain, Dominique. "The French Model of Urban Services." *West European Politics* 15, no. 2 (1992): 77–92.

———. "La firme locale-globale: Lyonnaise des Eaux (1980–2004)." *Sociologie du travail* 47, no. 3 (2005): 340–61. https://doi.org/10.4000/sdt.26667.

MacBride, Samantha. "Does Recycling Actually Conserve or Preserve Things?" *Discard Studies*, February 11, 2019. https://discardstudies.com/2019/02/11/12755/.

———. *Recycling Reconsidered the Present Failure and Future Promise of Environmental Action in the United States*. Cambridge, MA: MIT Press, 2012.

Mack, Adam. *Sensing Chicago: Noisemakers, Strikebreakers, and Muckrackers*. Urbana: University of Illinois Press, 2015.

Mackaman, Douglas. *Leisure Settings: Bourgeois Culture, Medicine, and the Spa in Modern France*. Chicago: University of Chicago Press, 1998.

Mallevialle, Joël, and I. H. Suffet, eds. *Identification and Treatment of Tastes and Odors in Drinking Water*. Denver, CO: American Water Works Association Research Foundation, 1987.

Malone, Charles T., and William H. McFadden. "Special Identification Detectors." In *Ancillary Techniques of Gas Chromatography*, edited by L. A. Ettre, 341–74. New York: Wiley, 1969.

Manheim, Marisa K., and Christy Spackman. "Embodied Rationality: A Framework of Human Action in Water Infrastructure Governance." *Current Opinion in Environmental Sustainability* 56 (June 1, 2022): 101170. https://doi.org/10.1016/j.cosust.2022.101170.

Martin, Emily. *Flexible Bodies: Tracking Immunity in American Culture from the Days of Polio to the Age of AIDS*. Boston: Beacon Press, 1994.

Marty, Nicolas. "La consommation des eaux embouteillées: Entre alimentation, distinction et hygiène." *Vingtième Siècle: Revue d'histoire* 3, no. 91 (2006): 25–41.

Mascha, Michael. *Fine Waters*. Philadelphia, PA: Quirk Books, 2006.

Masco, Joseph. *The Nuclear Borderlands: The Manhattan Project in Post–Cold War New Mexico*. Princeton, NJ: Princeton University Press, 2006.

Mayer, Harold M. "Politics and Land Use: The Indiana Shoreline of Lake Michigan." *Annals of the Association of American Geographers* 54, no. 4 (1964): 508–23.

McGuire, Michael. *The Chlorine Revolution*. Denver, CO: American Water Works Association, 2013.

McGuire, Michael J., Stuart W. Krasner, Cordelia J. Hwang, and George Izaguirre. "Closed-Loop Stripping Analysis as a Tool for Solving Taste and Odor Problems." *Journal AWWA* 73, no. 10 (1981): 530–37.

McGuire, Michael J., and Marie Pearthree. *Tucson Water Turnaround: Crisis to Success*. Denver, CO: American Water Works Association, 2020.

McMurray, Patrick. *Consider the Oyster: A Shucker's Field Guide*. New York: Thomas Dunne Books, 2007.

Meade, Mary. "Tastes Carried by Water Often Affect Cooking: Chemicals May Ruin the Flavor of Dishes." *Chicago Daily Tribune (1923–1963)*, April 15, 1934. ProQuest Historical Newspapers.

Meade, Nigel, and Towhidul Islam. "Modelling and Forecasting the Diffusion of Innovation—A 25-Year Review." *International Journal of Forecasting* 22, no. 3 (2006): 519–45. https://doi.org/10.1016/j.ijforecast.2006.01.005.

Medsker, Lloyd L., David Jenkins, and Jerome F. Thomas. "Odorous Compounds in Natural Waters: 2-exo-hydroxy-2-methylbornane, the Major Odorous Compound Produced by Several Actinomycetes," *Environmental Science and Technology* 3, no. 5 (1969):476–77.

Meilgaard, M. C., D. S. Reid, and K. A. Wyborski. "Reference Standards for Beer Flavor Terminology System." *American Society of Brewing Chemists Journal* 40, no. 4 (1982): 119–28.

Meiselman, Herbert L., and Howard G. Schutz. "History of Food Acceptance Research in the US Army." *Appetite* 40 (2003): 199–216.

Melosi, Martin V. *The Sanitary City: Environmental Services in Urban America from Colonial Times to the Present (Abridged Edition)*. Pittsburgh, PA: University of Pittsburgh Press, 2008. https://ebookcentral.proquest.com/lib/claremont/detail.action?docID=2038890.

Mertens, Mayli. "Liminal Innovation Practices: Questioning Three Common Assumptions in Responsible Innovation." *Journal of Responsible Innovation* 5, no. 3 (2018): 280–98. https://doi.org/10.1080/23299460.2018.1495031.

Millian, Mark, Tennyson, Patricia A., and Shane Snyder. "Model Communication Plans for Increasing Awareness and Fostering Acceptance of Direct Potable Reuse, Project 13-02." WateReuse Foundation, 2015.

Mintel. *Still and Sparkling Waters: U.S. 2023*. Chicago: Mintel Group Ltd., 2023.

Mitchell, John. "MWD Flooded with 'Muddy Water' Complaints." *L.A. Times*, October 14, 1979, sec. WS1.

Mohsenpour, Seyedeh Fatemeh, Sebastian Hennige, Nicholas Willoughby, Adebayo Adeloye, and Tony Gutierrez. "Integrating Micro-Algae into

Wastewater Treatment: A Review." *Science of the Total Environment* 752 (January 15, 2021): 142168. https://doi.org/10.1016/j.scitotenv.2020.142168.

Montoya, Theresa. "Permeable: Politics of Extraction and Exposure on the Navajo Nation." PhD diss., New York University, 2019.

Morrot, Gil, Frédéric Brochet, and Denis Dubourdieu. "The Color of Odors." *Brain and Language* 79, no. 2 (2001): 309–20. https://doi.org/10.1006/brln.2001.2493.

Mullins, Nicholas C. "The Development of a Scientific Specialty: The Phage Group and the Origins of Molecular Biology." *Minerva* 10, no. 1 (1972): 51–82.

Muniesa, Fabian, and Anne-Sophie Trébuchet-Breitwiller. "Becoming a Measuring Instrument." *Journal of Cultural Economy* 3, no. 3 (2010): 321–37. http://dx.doi.org/10.1080/17530350.2010.506318.

Murphy, Michelle. *Sick Building Syndrome and the Problem of Uncertainty: Environmental Politics, Technoscience, and Women Workers*. Durham, NC: Duke University Press, 2006.

National Research Council. *Water Reuse: Potential for Expanding the Nation's Water Supply Through Reuse of Municipal Wastewater*. Washington, DC: National Academies Press, 2012.

Nestlé Waters. "How to Taste Water," 2014. www.Nestlé-waters.com/get-to-know-us/through-our-waters/how-to-taste-water.

———. *L'Eau à la bouche/A Taste for Water*. Paris: Editions Textuel, 2009.

Noble, A. C., R. A. Arnold, B. M. Masuda, S. D. Pecore, J. O. Schmidt, and P. M. Stern. "Progress Towards a Standardized System of Wine Aroma Terminology." *American Journal of Enology and Viticulture* 35, no. 2 (1984): 107–9.

Norcom, G. D., and R. I. Dodd. "Activated Carbon for the Removal of Odor and Taste." *Journal (American Water Works Association)* 22 (1930): 1414–37.

Noss, Richard. "Distinguished Lecturer for 1987." *Association of Environmental Engineering Professors (AEEP) Newsletter* 21, no. 2 (1986): 2–3. www.aeesp.org/sites/default/files/newsletters/AEESPNL.21.2.1986.pdf.

O'Callaghan, Paul, Lakshmi Manjoosha Adapa, and Cees Buisman. "How Can Innovation Theories Be Applied to Water Technology Innovation?" *Journal of Cleaner Production* 276 (December 10, 2020): 122910. https://doi.org/10.1016/j.jclepro.2020.122910.

*OED Online*. Oxford University Press, December 2022. S.v. "Valorization, n." www-oed-com.ezproxy1.lib.asu.edu/view/Entry/221233?redirectedFrom=valorization.

Ofstehage, Andrew. "The Construction of an Alternative Quinoa Economy: Balancing Solidarity, Household Needs, and Profit in San Agustín, Bolivia." *Agriculture and Human Values* 29, no. 4 (2012): 441–54. https://doi.org/10.1007/s10460-012-9371-0.

Oregon Secretary of State. "Chapter 340, Division 55: Recycled Water Use." In *Oregon Administrative Rules Compilation for 2022*. Oregon Secretary of State, 2022. https://secure.sos.state.or.us/oard/displayDivisionRules.action?selectedDivision=1472.

Ormerod, Kerri Jean. "Common Sense Principles Governing Potable Water Recycling in the Southwestern US: Examining Subjectivity of Water Stewards Using Q Methodology." *Geoforum* 86 (November 1, 2017): 76–85. https://doi.org/10.1016/j.geoforum.2017.09.004.

———. "Illuminating Elimination: Public Perception and the Production of Potable Water Reuse." *WIREs: Water* 3, no. 4 (2016): 537–47. https://doi.org/10.1002/wat2.1149.

———. "Toilet Power: Potable Reuse and the Situated Meaning of Sustainability in the Southwestern United States." *Journal of Political Ecology* 26 (2019): 633–51.

Ortlieb, Stefan A., Werner A. Kügle, and Claus-Christian Carbon. "Fechner (1866): The Aesthetic Association Principle—A Commented Translation." *I-Perception* 11, no. 3 (2020): 1–20. https://doi.org/10.1177/2041669520920309.

Osborn, L. C. "Phenol Tastes in Chlorinated Water." *Journal AWWA* 17, no. 5 (1927): 586–90.

Pacheco-Vega, Raúl. "Human Right to Water and Bottled Water Consumption: Governing the Intersection of Water Justice, Rights and Ethics." In *Water Politics: Governance, Justice and the Right to Water*, edited by Farhana Sultana and Alex Loftus, 113–28. New York: Routledge, 2020.

———. "(Re)Theorizing the Politics of Bottled Water: Water Insecurity in the Context of Weak Regulatory Regimes." *Water* 1, no. 4 (2019): 658–74. https://doi.org/10.3390/w11040658.

Pangborn, Rose M., Harold W. Berg, and Brenda Hansen. "The Influence of Color on Discrimination of Sweetness in Dry Table-Wine." *American Journal of Psychology* 76, no. 3 (1963): 492–95. https://doi.org/10.2307/1419795.

Parinet, Julien, Manuel J. Rodriguez, and Jean Sérodes. "Influence of Water Quality on the Presence of Off-Flavour Compounds (Geosmin and

2-Methylisoborneol)." *Water Research* 44, no. 20 (2010): 5847–56. https://doi.org/10.1016/j.watres.2010.06.070.

Parker, Thomas. *Tasting French Terroir: The History of an Idea*. Berkeley: University of California Press, 2015.

Parr, Joy. "Local Water Diversely Known: Walkerton Ontario, 2000 and After." *Environment and Planning D: Society and Space* 23, no. 2 (2005): 251–71. https://doi.org/10.1068/d431.

———. *Sensing Changes: Technologies, Environments, and the Everyday 1953–2003*. Vancouver: University of British Columbia Press, 2010.

———. "What Makes Washday Less Blue? Gender, Nation, and Technology Choice in Postwar Canada." *Technology and Culture* 38, no. 1 (1997): 153–86. https://doi.org/10.2307/3106787.

Parrinello, Giacomo, Etienne S. Benson, and Wilko Graf von Hardenberg. "Estimated Truths: Water, Science, and the Politics of Approximation." *Journal of Historical Geography* 68 (April 1, 2020): 3–10. https://doi.org/10.1016/j.jhg.2020.03.006.

Parry, Alexander I. "Delivering Bacteriology to the American Homemaker: Correspondence Education, Kitchen Experiments, and Public Health, 1890–1930." *Isis* 114, no. 2 (2023): 317–40. https://doi.org/10.1086/725048.

Pavenello, Renato. "The International Association on Water Pollution Research." *Environmental Conservation* 1, no. 4 (1974): 280. http://dx.doi.org/10.1017/S0376892900004896.

Paxson, Heather. *The Life of Cheese: Crafting Food and Value in America*. Berkeley: University of California Press, 2012.

Pederson, G. L., Thomas H. Yorke, and J. K. Stamer. *Work Plan for the Schuylkill River Basin, Pennsylvania: Assessment of River Quality as Related to the Distribution and Transport of Trace Metals and Organic Substances*. Washington, DC: US Geological Survey, 1980.

Pérez-González, A., A. M. Urtiaga, R. Ibáñez, and I. Ortiz. "State of the Art and Review on the Treatment Technologies of Water Reverse Osmosis Concentrates." *Water Research* 46, no. 2 (2012): 267–83. https://doi.org/10.1016/j.watres.2011.10.046.

———. "Sensory Properties and Analysis of Two Muddy Odour Compounds, Geosmin and 2-Methylisoborneol, in Water and Fish." *Water Research* 14, no. 8 (1980): 1113–18. https://doi.org/10.1016/0043-1354(80)90161-X.

Peryam, David R. "Field Testing of Armed Forces Rations." In *Food Acceptance Testing Methodology: A Symposium Sponsored by the U.S. Quartermaster Food and Container Institute for the Armed Forces, Chicago, and the Quartermaster Research and Development Command U.S. Army Quartermaster Corps*, edited by David R. Peryam, Francis J. Pilgrim, and Martin S. Peterson, 75–85. Washington, DC: Advisory Board on Quartermaster Research and Development Committee on Foods, National Academy of Sciences, National Research Council, 1954.

Peryam, David R., Francis J. Pilgrim, and Martin S. Peterson, eds. *Food Acceptance Testing Methodology: A Symposium Sponsored by the U.S. Quartermaster Food and Container Institute for the Armed Forces, Chicago, and the Quartermaster Research and Development Command U.S. Army Quartermaster Corps*. Washington, DC: Advisory Board on Quartermaster Research and Development Committee on Foods, National Academy of Sciences, National Research Council, 1954.

Petrick, Gabriella M. "Feeding the Masses: H. J. Heinz and the Creation of Industrial Food." *Endeavour* 33, no. 1 (2009): 29–34.

Peynaud, Émile. *The Taste of Wine: The Art and Science of Wine Appreciation*. New York: Wiley, 1996.

Pezon, Christelle. *Le Service d'eau potable en France de 1850 à 1995*. Paris: Conservatoire National des Arts et Métiers/Presses du CEREM, , 2000. https://shs.hal.science/halshs-02549509/document.

Phelps, Earle B. "The Chemical Measure of Stream Pollution and Specification for Sewage Effluents." *American Journal of Public Health* 2 (1912): 524–34.

Phillips, Christopher J. "The Taste Machine: Sense, Subjectivity, and Statistics in the California Wine World." *Social Studies of Science* 46, no. 3 (2016): 461–68. https://doi.org/10.1 177/0306312716651504.

*Pima County FYI External Newsletter*. "County-Led Effort Finds New Ways to Brew up Reclaimed Water." September 15, 2017. https://webcms.pima.gov/cms/One.aspx?portalId=169&pageId=369008.

Pine, Kathleen H., and Max Liboiron. "The Politics of Measurement and Action." In *Proceedings of the 33rd Annual ACM Conference on Human Factors in Computing Systems (ACM)*. Seoul, Republic of Korea, 2015.

Pippen, Eldon L. "Poultry Flavor." In *Symposium on Foods: The Chemistry and Physiology of Flavors*, edited by H. W. Schultz, E. A. Day, and L. M. Libbey, ch. 11. Westport, CT: Avi Publishing, 1967.

Piriou, P., R. Devesa, M. De Lalande, and K. Glucina. "European Reassessment of MIB and Geosmin Perception in Drinking Water." *Journal of Water Supply: Research and Technology-Aqua* 58, no. 8 (2009): 532–38. https://doi.org/10.2166/aqua.2009.124.

Pixelis. "Eau de Paris, Ouvrez un grand cru." Dailymotion, August 6, 2013. Video. www.dailymotion.com/video/x12rhxb.

Pomranz, Mike. "Wastewater Beer's Sobering Moral: Many Still Don't Understand Recycled Water." *Food & Wine*, October 6, 2017. www.foodandwine.com/beer/wastewater-beers-sobering-moral-many-still-dont-understand-recycled-water.

Poncet, André. "La Loi du 16 décembre 1964 relative au régime et à la répartition des eaux et à la lutte contre leur pollution." *Revue forestière française* 5 (May 1965): 333–40.

Proctor, Robert. "Agnotology: A Missing Term to Describe the Cultural Production of Ignorance (and Its Study)." In *Agnotology: The Making and Unmaking of Ignorance*, edited by Robert Proctor and Londa Schiebinger, 1–36. Palo Alto, CA: Stanford University Press, 2008.

Proctor, Robert N., and Londa Schiebinger, eds. *Agnotology: The Making and Unmaking of Ignorance*. Palo Alto, CA: Stanford University Press, 2008.

Profita, Cassandra. "Why Dump Treated Wastewater When You Could Make Beer with It?" *NPR*, January 28, 2015. www.npr.org/sections/thesalt/2015/01/28/381920192/why-dump-treated-wastewater-when-you-could-make-beer-with-it.

*Progress on Household Drinking Water, Sanitation and Hygiene 2000–2020: Five Years into the SDGs*. Geneva: World Health Organization (WHO) and the United Nations Children's Fund (UNICEF), 2021.

Puget, Sabine. "Chlorine Flavour Perception and Neutralization in Drinking Water." University of Bourgogne, 2010. https://tel.archives-ouvertes.fr/tel-00786522/document.

Puig de la Bellacasa, María. "Encountering Bioinfrastructure: Ecological Struggles and the Sciences of Soil." *Social Epistemology* 28, no. 1 (2014): 26–40. https://doi.org/10.1080/02691728.2013.862879.

Pulido, Laura. "Flint, Environmental Racism, and Racial Capitalism." *Capitalism Nature Socialism* 27, no. 3 (2016): 1–16. https://doi.org/10.1080/10455752.2016.1213013.

Purnell, Howard. *Gas Chromatography*. New York: John Wiley & Sons, 1962.

Raab, Frank. "Taste and Odor Troubles in the Minneapolis Water Supply." *Journal AWWA* 23, no. 3 (1931): 430–34.

Rappert, Brian, and Wenda K. Bauchspies. "Introducing Absence." *Social Epistemology* 28, no. 1 (2014): 1–3. https://doi.org/10.1080/02691728.2013.862875.

Raspet, Sean. "Formulation 0.10." KW Institute for Contemporary Art, 2016. http://bb9.berlinbiennale.de/formulation-0-10/.

Ratcliffe, Rebecca. "Singapore Craft Beer Uses Recycled Sewage to Highlight Water Scarcity." *Guardian*, July 1, 2022. www.theguardian.com/world/2022/jul/01/singapore-craft-beer-newbrew-uses-recycled-sewage-highlight-water-scarcity.

Ray, Krishnendu. "Bringing the Immigrant Back into the Sociology of Taste." *Appetite* 119 (December 1, 2017): 41–47. https://doi.org/10.1016/j.appet.2016.10.013.

Reed, Atavia. "TikTok's Dilla the Urban Historian Is Educating Followers and Challenging Chicago Stereotypes—60 Seconds at a Time." Block Club Chicago, May 20, 2021. https://blockclubchicago.org/2021/05/20/tiktoks-dilla-the-urban-historian-is-educating-followers-and-challenging-chicago-stereotypes-60-seconds-at-a-time/.

"Report of an Investigation of the Pollution of Lake Michigan in the Vicinity of South Chicago and Indiana Harbors." *Public Health Reports (1896–1970)* 42, no. 35 (1927): 2200–2202.

Rice, Jacelyn, Amber Wutich, and Paul Westerhoff. "Assessment of De Facto Wastewater Reuse across the U.S.: Trends between 1980 and 2008." *Environmental Science & Technology* 47, no. 19 (2013): 11099–105. https://doi.org/10.1021/es402792s.

Richards, Ellen H., and J.W. Ellms. "The Coloring Matter of Natural Waters, Its Source, Composition, and Quantitative Measurement." *Journal of the American Chemical Society* 18, no. 1 (1896): 68–81.

Roberts, Lissa. "The Death of the Sensuous Chemist: The 'New' Chemistry and the Transformation of Sensuous Technology." *Studies in History and Philosophy of Science Part A* 26, no. 4 (December 1995): 503–29. https://doi.org/10.1016/0039-3681(95)00013-5.

Robison, Jason A., and Douglas S. Kenney. "Equity and the Colorado River Compact." *Environmental Law* 42, no. 4 (2012): 1157–1209.

Roccaro, Paolo. "Treatment Processes for Municipal Wastewater Reclamation: The Challenges of Emerging Contaminants and Direct Potable Reuse." *Current Opinion in Environmental Science & Health* 2 (April 1, 2018): 46–54. https://doi.org/10.1016/j.coesh.2018.02.003.

Rogers, Everett M. *Diffusion of Innovations*. 5th ed. Riverside: Free Press, 2003. ProQuest Ebook Central.

Romeyn, Travis R., Wesley Harijanto, Sofia Sandoval, Saied Delagah, and Mohamadali Sharbatmaleki. "Contaminants of Emerging Concern in Reverse Osmosis Brine Concentrate from Indirect/Direct Water Reuse Applications." *Water Science and Technology* 73, no. 2 (2015): 236–50. https://doi.org/10.2166/wst.2015.480.

Roosth, Sophia. "Of Foams and Formalisms: Scientific Expertise and Craft Practice in Molecular Gastronomy." *American Anthropologist* 115, no. 1 (2013): 4–16. https://doi.org/10.1111/j.1548-1433.2012.01531.x.

Rose, Nikolas. "Molecular Biopolitics, Somatic Ethics and the Spirit of Biocapital." *Social Theory & Health* 5 (2007): 3–29. https://doi-org.ezproxy1.lib.asu.edu/10.1057/palgrave.sth.8700084.

———. "The Politics of Life Itself." *Theory, Culture & Society* 18, no. 6 (2001): 1–30. https://doi.org/10.1177/02632760122052020.

Rosen, A. A., C. I. Mashni, and R. S. Safferman. "Recent Developments in the Chemistry of Odour in Water: The Cause of Earthy/Musty Odour." *Water Treatment* 19 (1970): 106–19.

Rosenberg, Daniel. "Data before the Fact." In *"Raw Data" Is an Oxymoron*, edited by Lisa Gittleman, 15–40. Cambridge, MA: MIT Press, 2013.

Roy, Siddhartha, and Marc A. Edwards. "Preventing Another Lead (Pb) in Drinking Water Crisis: Lessons from the Washington D.C. and Flint MI Contamination Events." In "Drinking Water Contaminants," edited by Susan Richardson and Cristina Postigo, special issue, *Current Opinion in Environmental Science & Health* 7 (February 1, 2019): 34–44. https://doi.org/10.1016/j.coesh.2018.10.002.

Royte, Elizabeth. *Bottlemania: How Water Went on Sale & Why We Bought It*. London: Bloomsbury, 2008.

———. "A Tall, Cool Drink of . . . Sewage?" *New York Times Magazine*, August 8, 2008. www.nytimes.com/2008/08/10/magazine/10wastewater-t.html.

Saito, Yuriko. "Everyday Aesthetics." *Philosophy and Literature* 25 (2001): 87–95.

Sayre, Ida M. "The French Connection: Cuisine and Water Treatment." *Journal AWWA* 78, no. 7 (1986): 52–61.

Schaffartzik, Anke, Andreas Mayer, Simone Gingrich, Nina Eisenmenger, Christian Loy, and Fridolin Krausmann. "The Global Metabolic Transition: Regional Patterns and Trends of Global Material Flows, 1950–2010." *Global Environmental Change* 26 (May 1, 2014): 87–97. https://doi.org/10.1016/j.gloenvcha.2014.03.013.

Schiebinger, Londa. "Feminist History of Colonial Science." *Hypatia* 19, no. 1 (2004): 233–54.

Schmidt, Charles W. "The Yuck Factor When Disgust Meets Discovery." *Environmental Health Perspectives* 116, no. 12 (2008): A524–27. https://doi.org/10.1289/ehp.116-a524.

Schneider, D. *Hybrid Nature: Sewage Treatment and the Contradictions of the Industrial Ecosystem*. Cambridge, MA: MIT Press, 2011.

Schneier-Madanes, Graciela. *Globalized Water: A Question of Governance*. New York: Springer, 2014.

Scholes, Rachel C., Angela N. Stiegler, Cayla M. Anderson, and David L. Sedlak. "Enabling Water Reuse by Treatment of Reverse Osmosis Concentrate: The Promise of Constructed Wetlands." *ACS Environmental Au* 1, no. 1 (2021): 7–17. https://doi.org/10.1021/acsenvironau.1c00013.

Schulhof, Pierre. "Water Supply in the Paris Suburbs: Changing Treatment for Changing Demands." *Journal AWWA* 72, no. 8 (1980): 428–34. https://doi.org/10.1002/j.1551-8833.1980.tb04549.x.

Schummer, Joachim. "The Impact of Instrumentation on Chemical Species Identity: From Chemical Substances to Molecular Species." In *From Classical to Modern Chemistry: The Instrumental Revolution*, 188–211. Cambridge, UK: Royal Society of Chemistry, 2002.

Scottsdale Arizona. "One Water Brewing Showcase." July 15, 2020. YouTube video. www.youtube.com/watch?v=gt5sldaElTo.

Scottsdale Water. "One Water Brewing Showcase." 2019. http://canalconvergence.com/wp-content/uploads/2019/07/one_water_brewing_showcase_2019.pdf.

Scruggs, Caroline E., Claudia B. Pratesi, and John R. Fleck. "Direct Potable Water Reuse in Five Arid Inland Communities: An Analysis of Factors Influencing Public Acceptance." *Journal of Environmental Planning and Management* 63, no. 8 (2020): 1470–1500. https://doi.org/10.1080/09640568.2019.1671815.

Selin, Cynthia. "The Sociology of the Future: Tracing Stories of Technology and Time." *Sociology Compass* 2, no. 6 (2008): 1878–95. https://doi.org/10.1111/j.1751-9020.2008.00147.x.

Selin, Cynthia, and Jathan Sadowski. "Against Blank-Slate Futuring." In *Remaking Participation: Science, Environment and Emerging Publics*, edited by Jason Chilvers and Matthew Kearnes, 218–37. New Yor: Routledge, 2016.

Shapin, Steven. *Changing Tastes: How Foods Tasted in the Early Modern Period and How They Taste Now.* The Hans Rausing Lecture, Salvia Småskrifter 14. Uppsala: Tryck Wikströms, for the University of Uppsala, 2011.

———. "The Sciences of Subjectivity." *Social Studies of Science* 42, no. 2 (2012): 170–84. https://doi.org/10.1177/0306312711435375.

———. "A Taste of Science: Making the Subjective Objective in the California Wine World." *Social Studies of Science* 46 (2016): 436–60.

Shapiro, Laura. *Perfection Salad: Women and Cooking at the Turn of the Century.* Berkeley: University of California Press, 2008.

Shapiro, Nicholas. "Attuning to the Chemosphere: Domestic Formaldehyde, Bodily Reasoning, and the Chemical Sublime." *Cultural Anthropology* 30, no. 3 (2015): 368–93.

Shapiro-Shapin, Carolyn G. "Filtering the City's Image: Progressivism, Local Control, and the St. Louis Water Supply, 1890–1906." *Journal of the History of Medicine* 54 (July 1999): 387–412.

Sheppard, Stephen R. J., Alison Shaw, David Flanders, Sarah Burch, Arnim Wiek, Jeff Carmichael, John Robinson, and Stewart Cohen. "Future Visioning of Local Climate Change: A Framework for Community Engagement and Planning with Scenarios and Visualisation." In "Community Engagement for Sustainable Urban Futures," edited by Greg Hearn, Marcus Foth, and Tony Stevenson, special issue, *Futures* 43, no. 4 (2011): 400–412. https://doi.org/10.1016/j.futures.2011.01.009.

Shesgreen. "Wet Dogs and Gushing Oranges: Winespeak for a New Millennium." *Chronicle of Higher Education*, March 7, 2003. www.chronicle.com/article/wet-dogs-and-gushing-oranges-winespeak-for-a-new-millennium/.

Shields-Argèles, Christy. "A Cooperative Model of Tasting: Comté Cheese and the Jury Terroir." *Food, Culture, and Society* 22, no. 2 (2019): 168–85.

Silvey, J. K. G. "Relation of Irrigation Runoff to Tastes and Odors." *Journal (American Water Works Association)* 45, no. 11 (November 1953): 1179–86.

Sinclair, Upton. *The Jungle*. Minneapolis, MN: Lerner, 1906.

Sjöström, Loren B., and Stanley Cairncross. "What Makes Flavor Leadership?" *Food Technology* 7, no. 2 (February 1953): 56–58.

Slaff, Steven. "Land Subsidence and Earth Fissures in Arizona." Down to Earth Series 3. Arizona Geological Survey, 1993. https://repository.arizona.edu/bitstream/handle/10150/629605/dte-3_subsidence_1993.pdf?sequence=1&isAllowed=y.

Smith, Carl. *City Water, City Life: Water and the Infrastructure of Ideas in Urbanizing Philadelphia, Boston, and Chicago*. Chicago: University of Chicago Press, 2013.

Smith, H. M., S. Brouwer, P. Jeffrey, and J. Frijns. "Public Responses to Water Reuse—Understanding the Evidence." *Journal of Environmental Management* 207 (February 1, 2018): 43–50. https://doi.org/10.1016/j.jenvman.2017.11.021.

Smith, Harold. "South Side Gets 125 Acre Park: Site of Water Filter Will Be a Beauty Spot." *Chicago Daily Tribune (1923–1963)*, October 9, 1941, S1. ProQuest Historical Newspapers.

Smith, Lindsay A. "The Missing, the Martyred and the Disappeared: Global Networks, Technical Intensification and the End of Human Rights Genetics." *Social Studies of Science* 47, no. 3 (2017): 398–416. https://doi.org/10.1177/0306312716678489.

Smith, Mark M. *A Sensory History Manifesto*. University Park: Pennsylvania State University Press, 2021.

Smith-Howard, Kendra. *Pure and Modern Milk: An Environmental History since 1900*. Oxford: Oxford University Press, 2013.

Smits, Fenna, and Rebeca Ibáñez Martín. "'The Village' as a Site for Multispecies Innovation: Rethinking the Village in Response to the Anthropocene." *Etnofoor* 31, no. 2 (2019): 67–86.

Snyder, Shane, Paul Westerhoff, Yeomin Yoon, and David Sedlak. "Pharmaceuticals, Personal Care Products, and Endocrine Disruptors in Water: Implications for the Water Industry." *Environmental Engineering Science* 20, no. 5 (2003): 449–69.

Spackman, Christy. "In Smell's Shadow: Materials and Politics at the Edge of Perception." *Social Studies of Science* 50, no. 3 (2020): 418–39. https://doi.org/10.1177/0306312720918946.

———. "Just Noticeable: Erasing Place in Municipal Water Treatment in the U.S. during the Interwar Period." *Journal of Historical Geography* 67 (January 1, 2020): 2–13. https://doi.org/10.1016/j.jhg.2019.10.014.

———. "Ordering Volatile Openings: Instrumentation and the Rationalization of Bodily Odors." *Food, Culture & Society* 22, no. 5 (2019): 674–91. https://doi.org/10.1080/15528014.2019.1638135.

———. "Transforming Taste: The Twentieth-Century Aesthetic Remaking of Water." PhD diss., New York University, 2015.

Spackman, Christy, and Gary Burlingame. "Sensory Politics: The Tug-of-War between Potability and Palatability in Municipal Water Production." *Social Studies of Science* 48, no. 3 (June 2018): 350–71.

Spackman, Christy, and Jacob Lahne. "Sensory Labor: Considering the Work of Taste in the Food System." *Food, Culture, and Society* 22, no. 2 (2019): 142–51.

Spackman, Christy, Marisa K. Manheim, and Shomit Barua. "Tasting Water at Canal Convergence 2021: An Experiment in Embodied Remembering." *Gastronomica* 22, no. 4 (2022): 54–70.

Spalding, G. R. "Activated Char as a Deodorant in Water Treatment." *Journal (American Water Works Association)* 22 (1930): 646–48.

Spary, E. C. *Feeding France: New Sciences of Food, 1760-1815*. Cambridge: Cambridge University Press, 2014.

Spaulding, Charles H. "Accuracy and Application of Threshold Odor Tests." *Journal (American Water Works Association)* 34, no. 6 (1942): 877–910.

———. "Preammoniation at Springfield, Illinois." *Journal (American Water Works Association)* 21, no. 8 (1929): 1085–96.

———. "Quantitative Determination of Odor in Water." *American Journal of Public Health* 21 (September 1931): 1038–39.

———. "Some Quantitative Odor Determinations." *Journal (American Water Works Association)* 24, no. 8 (1932): 1111–18.

Steel, Ernest W. "By-Products from Industrial Wastes." *Scientific American* 143, no. 5 (1930): 378–79.

Stewart, George F. "The Challenge in Flavor Research." *Food Technology*, January 1963, 5.

———. "Effect of Nutrition on Quality of Poultry Meat and Eggs." *World's Poultry Science Journal* 8, no. 4 (1952): 246–48. https://doi.org/10.1079/WPS19520040.

Stewart, Ollie. "Report from Europe: Finds Water in Paris Okay." *Baltimore Afro-American*, March 26, 1960. ProQuest Historical Newspapers.

Stoll, Mark. "Rachel Carson's *Silent Spring*: A Book That Changed the World." *Environment and Society Portal Virtual Exhibitions* 1 (2007). http://www.environmentandsociety.org/exhibitions/silent-spring/silent-spring-international-best-seller.

Stone, Herbert, Rebecca N. Bleibaum, and Heather A. Thomas. "Introduction to Sensory Evaluation." In *Sensory Evaluation Practices*, 4th ed. Oxford: Academic Press, 2012.

Stradling, David, and Richard Stradling. "Perceptions of the Burning River: Deindustrialization and Cleveland's Cuyahoga River." *Environmental History* 13, no. 3 (2008): 515–35. https://doi.org/10.1093/envhis/13.3.515.

Stratton, G. M. "Review of *The Applicability of Weber's Law to Smell*." *Psychological Review* 6, no. 5 (1899): 557–59. https://doi.org/10.1037/h0067166.

Stratton-Childers, LaShell. "More Beer, Less Yuck Factor." *Water Environment & Technology* 27, no. 5 (2015): 23–24, 26.

Streicher, Lee. "Colorado River Aqueduct System." *Journal (American Water Works Association)* 50, no. 9 (1958): 1223–26.

Suffet, I. H. (Mel), S. Braithwaite, Y. Zhou, and A. Bruchet. "The Drinking Water Taste-and-Odour Wheel after 30 Years." In *Taste and Odour in Source and Drinking Water: Causes, Controls, and Consequences*, edited by T.-F. Lin, S. Watson, A. M. Dietrich, and I. H. (Mel) Suffet, 11–46. London: IWA Publishing, 2019.

Suffet, I. H., and S. Segall. "Detecting Taste and Odor in Drinking Water." *Journal (American Water Works Association)* 63, no. 9 (1971): 608.

Summitt, April R. *Contested Waters: An Environmental History of the Colorado River*. Boulder: University of Colorado Press, 2013.

Swyngedouw, Erik. "Dispossessing H2O: The Contested Terrain of Water Privatization." *Capitalism Nature Socialism* 16, no. 1 (2005): 81–98. https://doi.org/10.1080/1045575052000335384.

———. *Social Power and the Urbanization of Water: Flows of Power*. Oxford Geographical and Environmental Studies. Oxford: Oxford University Press, 2004.

Swyngedouw, Erik, and Rutgerd Boelens. "'. . . And Not a Single Injustice Remains:' Hydro-Territorial Colonization and Techno-Political Transformations in Spain." In *Water Justice*, edited by Rutgerd Boelens and J. Vos, 115–33. Cambridge: Cambridge University Press, 2018.

Tarr, Joel A. "Industrial Wastes and Public Health: Some Historical Notes, Part I, 1876–1932." *American Journal of Public Health* 75, no. 9 (1985): 1059–67.

———. "Searching for a 'Sink' for an Industrial Waste: Iron-Making Fuels and the Environment." *Environmental History Review* 18, no. 1 (1994): 9–34. https://doi.org/10.2307/3984743.

Tarr, Joel A., James McCurley, Francis C. McMichael, and Terry Yosie. "Water and Wastes: A Retrospective Assessment of Wastewater Technology in the United States, 1800–1932." *Technology and Culture* 25, no. 2 (1984): 226–63. https://doi.org/10.2307/3104713.

Teigen de Master, Kathryn, James LaChance, Sarah Bowen, and Lillian MacNell. "Terroir in Transition: Environmental Change in the Wisconsin Artisanal Cheese and New England Oyster Sectors." *Sustainability* 11, no. 10 (2019): 2969. https://doi.org/10.3390/su11102969.

Teil, Geneviève, and Antoine Hennion. "Discovering Quality or Performing Taste? A Sociology of the Amateur." In *Qualities of Food*, edited by Mark Harvey, Andrew McMeekin, and Alan Warde, 19–37. Manchester: Manchester University Press, 2004.

Teillet, Eric, Christine Urbano, Sylvie Cordelle, and Pascal Schlich. "Consumer Perception and Preference of Bottled and Tap Water." *Journal of Sensory Studies* 25, no. 3 (2010): 463–80. https://doi.org/10.1111/j.1745-459X.2010.00280.x.

Tenney, Warren. "5 Things You Need to Know Right Now about Arizona's Drought." AMWUA, June 18, 2018. www.amwua.org/blog/5-things-you-need-to-know-right-now-about-arizonas-drought-.

Tennyson, Patricia A., Mark Millan, and David Metz. "Getting Past the 'Yuck Factor': Public Opinion Research Provides Guidance for Successful Potable Reuse Outreach." *Journal AWWA* 107, no. 11 (2015): 58–62. https://doi.org/10.5942/jawwa.2015.107.0163.

Thaysen, A. C. "The Origin of an Earthy or Muddy Taint in Fish I: The Nature and Isolation of the Taint." *Annuals of Applied Biology* 23 (1936): 99—104.

Thomas, Harold A., Jr. "Calculation of Threshold Odor." *Journal AWWA* 35, no. 6 (1943): 751–69.

Thomas, Norton A. "Taste and Odor Control on Lake Michigan." *Journal (American Water Works Association)* 32, no. 7 (1940): 1183–86.

Thompson, Emily. *The Soundscapes of Modernity: Architectural Acoustics and the Culture of Listening in America, 1900–1933*. Cambridge, MA: MIT Press, 2002.

Tisdale, E. S., Norman F. Prince, Chairman Howard, Wellington Donaldson, James M. Caird, and C. R. Cox. "Cooperative State Control of Phenol Wastes on the Ohio River Watershed [with Discussion]." *Journal (American Water Works Association)* 18, no. 5 (1927): 574–86.

Tomes, Nancy. *The Gospel of Germs: Men, Women, and the Microbe in American Life*. Cambridge, MA: Harvard University Press, 1999.

Tompkins, Kyla Wazana. "Sylvester Graham's Imperial Dietetics." *Gastronomica* 9, no. 1 (2009): 50–60. https://doi.org/10.1525/gfc.2009.9.1.50.

Tracy, Sarah E. "Delicious: A History of Monosodium Glutamate and Umami, the Fifth Taste Sensation." PhD diss., University of Toronto, 2016.

———. "Delicious Molecules: Big Food Science, the Chemosenses, and Umami." *Senses and Society* 13, no. 1 (March 2018): 89–107.

Trubek, Amy. *The Taste of Place: A Cultural Journey into Terroir.* Berkeley: University of California Press, 2008.

Trubek, Amy, Kolleen M. Guy, and Sarah Bowen. "Terroir: A French Conversation with a Transnational Future." *Contemporary French and Francophone Studies* 14, no. 2 (2010): 139–48. https://doi.org/10.1080/17409291003644206.

Tullett, William. *Smell in Eighteenth-Century England: A Social Sense.* Oxford: Oxford University Press, 2019.

Twilley, Nicola. "Exploring Aeroir, or the Atmospheric Taste of Place." In *Food and Landscape: Proceedings of the 2017 Oxford Symposium on Food and Cookery*, edited by Mark McWilliams, 50–56. London: Prospect Books, 2018.

Ulloa, Ana María. "The Chef and the Flavorist: Reflections of the Value of Sensory Expertise." *Food, Culture, and Society* 22, no. 2 (2019): 186–202.

UNESCO World Water Assessment Programme. *The United Nations World Water Development Report 2020: Water and Climate Change.* 2020. UNESCO Digital Library. https://unesdoc.unesco.org/ark:/48223/pf0000372985.locale=en.

US Environmental Protection Agency (EPA), Office of Administration. "Basic Information on PFAS." Overviews and Factsheets. March 30, 2016. www.epa.gov/pfas/basic-information-pfas.

———. "EPA Issues Final List of Contaminants for Potential Regulatory Consideration in Drinking Water, Significantly Increases PFAS Chemicals for Review." News Release, November 2, 2022. www.epa.gov/newsreleases/epa-issues-final-list-contaminants-potential-regulatory-consideration-drinking-water.

US Environmental Protection Agency (EPA), Office of Policy. "History of the Clean Water Act." Overviews and Factsheets, February 22, 2013. www.epa.gov/laws-regulations/history-clean-water-act.

US Environmental Protection Agency (EPA), Office of Water. "Contaminants of Emerging Concern Including Pharmaceuticals and Personal Care Products." Reports and Assessments, August 18, 2015. www.epa.gov/wqc/contaminants-emerging-concern-including-pharmaceuticals-and-personal-care-products.

———. "Factoids: Drinking Water and Ground Water Statistics for 2007." March 2008. https://nepis.epa.gov/Exe/ZyPDF.cgi/P100N2VG.PDF?Dockey=P100N2VG.PDF.

———. "Information about Public Water Systems." Collections and Lists, September 21, 2015. /www.epa.gov/dwreginfo/information-about-public-water-systems.

US Water Alliance. "One Water Roadmap: The Sustainable Management of Life's Most Essential Resource." 2016. http://uswateralliance.org/sites/uswateralliance.org/files/publications/Roadmap%20FINAL.pdf.

Valente, Thomas W., and Everett M. Rogers. "The Origins and Development of the Diffusion of Innovations Paradigm as an Example of Scientific Growth." *Science Communication* 16, no. 3 (1995): 242–73.

VanDerslice, James. "Drinking Water Infrastructure and Environmental Disparities: Evidence and Methodological Considerations." *American Journal of Public Health* 101, no. S1 (2011): S109–14. https://doi.org/10.2105/AJPH.2011.300189.

Vaughn, J. C. "Laboratory Control and Operating Experiences at the Hammond, Indiana, Filtration Plant." *Journal (American Water Works Association)* 31, no. 12 (1939): 2137–48.

Velasco, Carlos, Yunwen Tu, and Marianna Obrist. "Towards Multisensory Storytelling with Taste and Flavor." In *Proceedings of the 3rd International Workshop on Multisensory Approaches to Human-Food Interaction*, 1–7. MHFI'18. New York: Association for Computing Machinery, 2018. https://doi.org/10.1145/3279954.3279956.

Voß, Jan Peter, and Michael Guggenheim. "Making Taste Public: Industrialized Orders of Sensing and the Democratic Potential of Experimental Eating." *Politics and Governance* 7, no. 4 (2019): 224–36.

Waldby, David, and Suzanne Grendahl. "First Full Scale DPR Permit in Arizona for Demonstration and Education." Paper presented at the WateReuse Association Webcast, May 28, 2020. https://watereuse.org/wp-content/uploads/2020/05/AZ_Water_DPR_Webcast.pdf.

Walker, Rob. "Bottled Water Leaks Volume as Credit Crunch Bites." *Euromonitor International*, February 19, 2009.

Wang, Yun, Xiaojuan Ma, Qiong Luo, and Huamin Qu. "Data Edibilization: Representing Data with Food." In *Proceedings of the 2016 CHI Conference Extended Abstracts on Human Factors in Computing Systems*, 409–22. CHI EA '16. New York: Association for Computing Machinery, 2016. https://doi.org/10.1145/2851581.2892570.

Warren, Kenneth. *The American Steel Industry, 1850–1970: A Geographical Interpretation*. Pittsburgh: University of Pittsburgh Press, 1987.

Warren, R. M., and Edward Bartow. "Taste and Odor in Chlorinated Water." *Journal AWWA* 11, no. 4 (1924): 881–86.

Water Citizen. "Llllet's Get Ready to Reuuuuse!!!!! Sustainable Beer Smackdown at WEFTEC(R)!" *Water Citizen News* (blog), September 27, 2015. http://watercitizennews.com/4783/.

Water Strategies. "Turning Reuse Water into Beer: Pure Water Brew." *Municipal Water Leader*, April 2019. https://issuu.com/waterstrategies/docs/april19muniup/s/94350.

Weiss, Brad. *Real Pigs: Shifting Values in the Field of Local Pork*. Durham, NC: Duke University Press, 2018.

Weisz, George. "Spas, Mineral Waters, and Hydrological Science in Twentieth-Century France." *Isis* 92, no. 3 (2001): 451–83.

Wells, Madeline. "I Tried Beer Made from Recycled Wastewater from an SF High-Rise: It May Be the Future." SFGate.com, March 28, 2023. www.sfgate.com/food/article/sf-recycled-wastewater-beer-from-high-rise-17862774.php.Wendt, Lloyd, and Herman Kogan. *Big Bill of Chicago*. Evanston, IL: Northwestern University Press, 2005.

West, Harry. "Thinking Like a Cheese: Towards an Ecological Understanding of the Reproduction of Knowledge in Contemporary Artisan Cheese Making." In *Understanding Cultural Transmission in Anthropology: A Critical Synthesis*, edited by Roy Ellen, Stephen J. Lycett, and Sarah E. Johns, 320–45. Oxford: Berghahn Books, 2013.

Whelton, A. J., A. M. Dietrich, G. A. Burlingame, M. Schechs, and S. E. Duncan. "Minerals in Drinking Water: Impacts on Taste and Importance to Consumer Health." *Water Science and Technology: A Journal of the International Association on Water Pollution Research* 55, no. 5 (2007): 283–91. https://doi.org/10.2166/wst.2007.190.

Whelton, Andrew J., and Andrea M. Dietrich. "Relationship between Intensity, Concentration, and Temperature for Drinking Water Odorants." *Water Research* 38, no. 6 (2004): 1604–14. https://doi.org/10.1016/j.watres.2003.11.036.

"Where the Colorado River Crisis Is Hitting Home." National Public Radio, *Morning Edition*, September 22, 2022. www.npr.org/2022/09/22/1124150368/where-the-colorado-river-crisis-is-hitting-home.

Whipple, George. *The Value of Pure Water.* New York: John Wiley & Sons, 1907.

Whipple, George C. "The Observation of Odor as an Essential Part of Water Analysis." *Public Health Papers and Reports* 25 (1899): 587–93.

Wick, Emily L. "Flavor Update: One Opinion." *Food Technology*, December 1966, 43, 46, 48.

Wilk, Richard. "Bottled Water: A Pure Commodity in the Age of Branding." *Journal of Consumer Culture* 6, no. 3 (2006): 303–25. https://doi.org/10.1177%2F1469540506068681.

"William A. Evans." *American Journal of Public Health* 39 (August 1949): 1039–40.

Worley, Jennifer L., Andrea M. Dietrich, and Robert C. Hoehn. "Dechlorination Techniques to Improve Sensory Odor Testing of Geosmin and 2-MIB." *Journal AWWA* 95, no. 3 (2003): 109–17. https://doi.org/10.1002/j.1551-8833.2003.tb10319.x.

Young, W. F., H. Horth, R. Crane, T. Ogden, and M. Arnott. "Taste and Odour Threshold Concentrations of Potential Potable Water Contaminants." *Water Research* 30, no. 2 (1996): 331–40. https://doi.org/10.1016/0043-1354(95)00173-5.

Zellner, D. A., and M. A. Kautz. "Color Affects Perceived Odor Intensity." *Journal of Experimental Psychology. Human Perception and Performance* 16, no. 2 (1990): 391–97. https://doi.org/10.1037//0096-1523.16.2.391.

# Index

Founded in 1893,
UNIVERSITY OF CALIFORNIA PRESS
publishes bold, progressive books and journals on topics in the arts, humanities, social sciences, and natural sciences—with a focus on social justice issues—that inspire thought and action among readers worldwide.

The UC PRESS FOUNDATION
raises funds to uphold the press's vital role as an independent, nonprofit publisher, and receives philanthropic support from a wide range of individuals and institutions—and from committed readers like you. To learn more, visit ucpress.edu/supportus.

Milton Keynes UK
Ingram Content Group UK Ltd.
UKHW020733060524
442199UK00003B/32